# SPUNKY SCIENCE

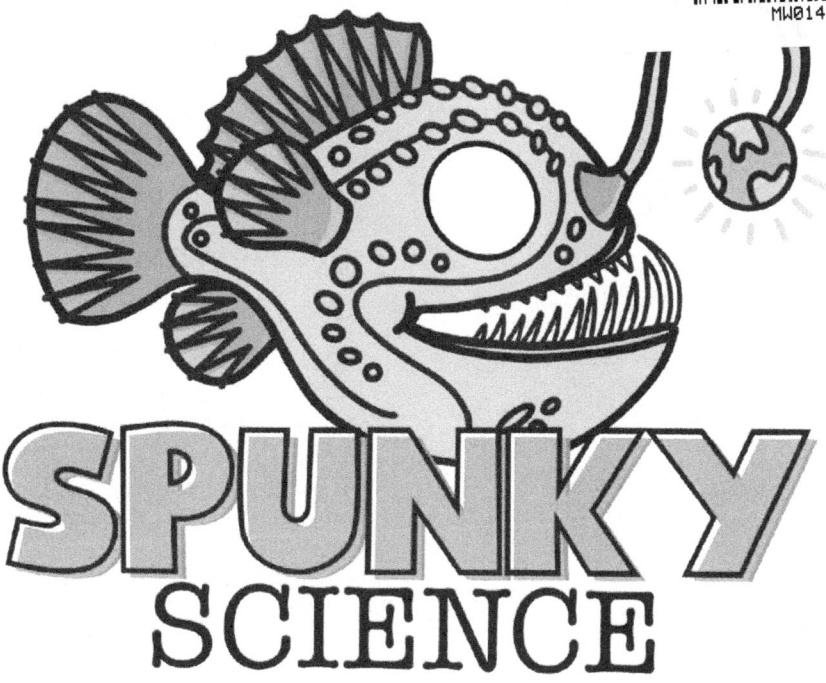

Copyright © 2022 Spunky Science

All rights reserved. No part of this book may be altered, reproduced, redistributed, or used in any manner other than its original intent without written permission or copyright owner except for the use of quotation in a book review.

## CAN:

✓ Make copies for your students for educational use

✓ Print content in different forms such as a booklet

✓ Print in various sizes to fit your needs

✓ Post content on a school-based platform for student use or reference

## CAN'T:

✗ Distribute digital or copies to others without an additional purchase

✗ Remove Spunky Science logo or copyright

✗ Resell or redistribute in any way other than originally intended by Spunky Science

# THIS BELONGS TO

_____

BIOLOGY

SPACE SCIENCE

PHYSICS

I love science

EARTH SCIENCE

CHEMISTRY

Spunky Science ©

# SAFETY

Hair tie for long hair

Splash goggles

Lab coat or apron

Closed-toe shoes

disposable gloves

eye wash station

fire extinguisher

FIRE BLANKET

## LAB SAFETY RULES:
- Do not eat or drink in the lab
- Wear appropriate PPE at all times
- Stay on task
- Follow lab instructions
- Dress appropriately
- Measure accurately
- Be careful when handling hot glassware
- Keep a clean work station
- Notify the teacher if glass breaks or equivalent

Sign here →

print here →

© SpunkyScience

# SCIENCE TOOLS

ruler and meter stick measure length

graduated cylinder

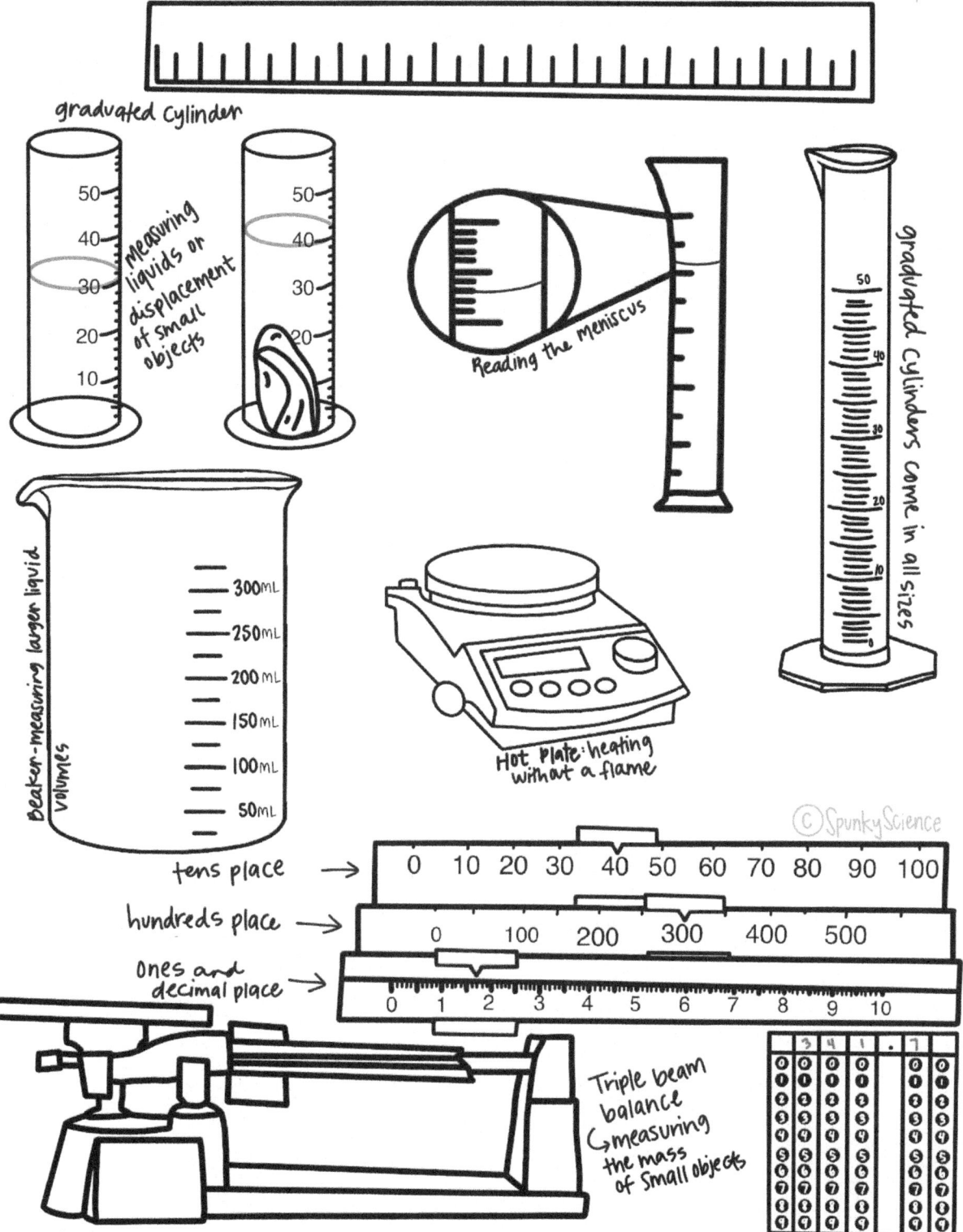

measuring liquids or displacement of small objects

Reading the meniscus

graduated cylinders come in all sizes

beaker - measuring larger liquid volumes

- 300 mL
- 250 mL
- 200 mL
- 150 mL
- 100 mL
- 50 mL

Hot plate: heating without a flame

tens place →
hundreds place →
ones and decimal place →

Triple beam balance ↳ measuring the mass of small objects

© Spunky Science

# SCIENTIFIC REASONING

**C**LAIM — A descriptive statement that answers the question- what side are you on?

**E**VIDENCE — Two or three facts or scientific data that directly supports your claim- why should we believe you?

**R**EASONING — A full explanation as to how your evidence supports the claim that you have made.

# PHENOMENA
## AN EXCEPTIONAL, UNUSUAL, OR ABNORMAL PERSON, EVENT, OR OCCURANCE.

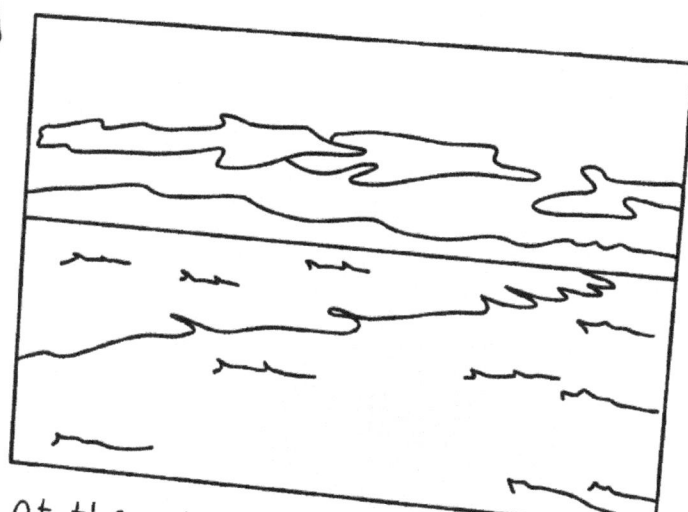

At the point where the Dead Sea and the Indian Sea meet, they do not mix. Why do you think this is happening?

What would happen if a hurricane collided with a tornado?

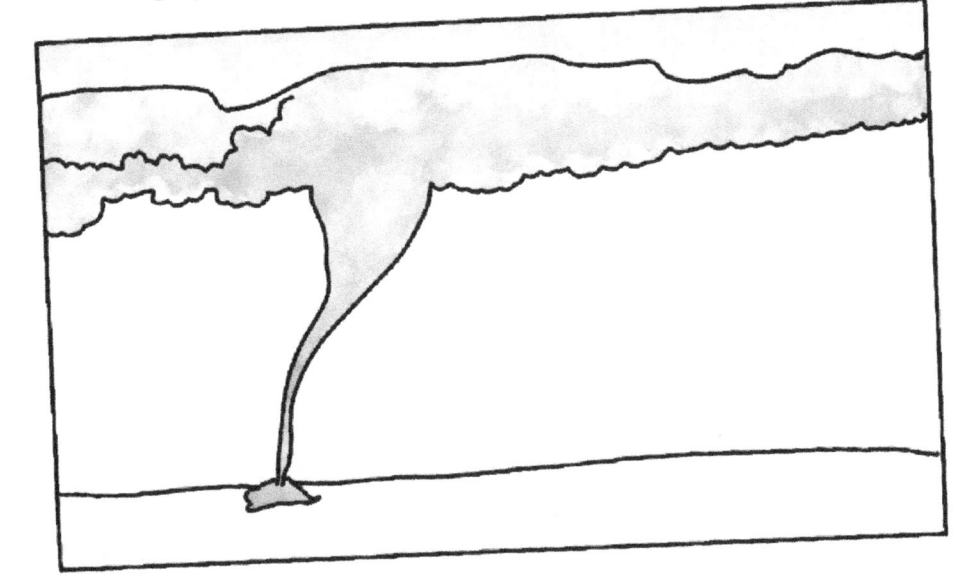

Why do our fingers get wrinkled in the water?

© SpunkyScience

# SCIENCE REFERENCE SHEET

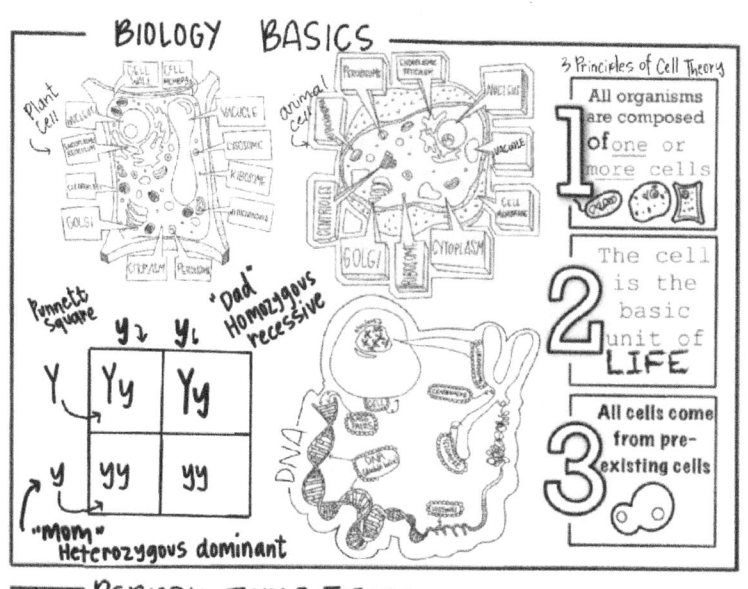

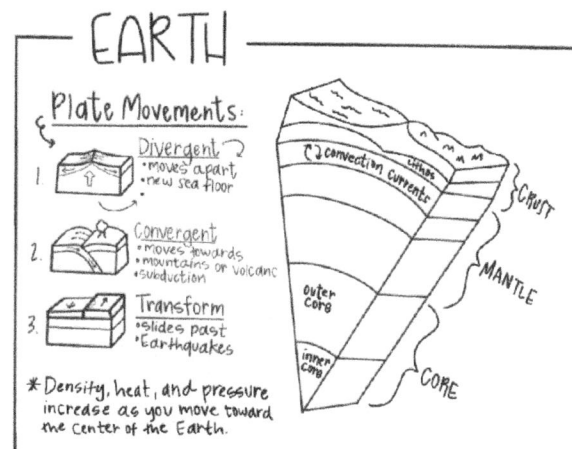

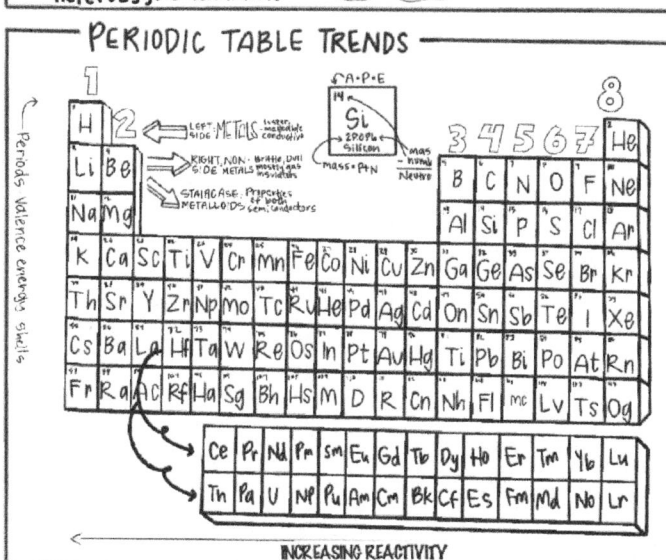

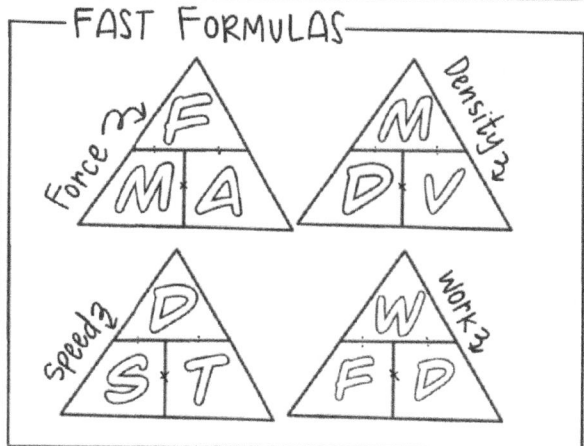

## BIOLOGY BASICS

Plant Cell / Animal Cell (labeled: CELL WALL, CELL MEMBRANE, NUCLEUS, VACUOLE, LYSOSOME, RIBOSOME, CHLOROPLAST, CENTRIOLES, GOLGI, CYTOPLASM, MITOCHONDRIA, ER)

DNA

### 3 Principles of Cell Theory
1. All organisms are composed of one or more cells
2. The cell is the basic unit of LIFE
3. All cells come from pre-existing cells

### Punnett Square

|   | Y  | Y  |
|---|----|----|
| y | Yy | Yy |
| y | yy | yy |

"Dad" — Homozygous recessive
"Mom" — Heterozygous dominant

## EARTH

### Plate Movements:
1. **Divergent** — moves apart, new sea floor
2. **Convergent** — moves towards, mountains or volcano, subduction
3. **Transform** — slides past, Earthquakes

* Density, heat, and pressure increase as you move toward the center of the Earth.

Earth layers: CRUST, MANTLE, outer CORE, inner CORE, convection currents

## PERIODIC TABLE TRENDS

LEFT SIDE METALS — luster, malleable, conductivity
RIGHT NON-METALS — brittle, dull, mostly gas, insulators
STAIRCASE METALLOIDS — properties of both, semiconductors

Si — A·P·E, 2 protons, Silicon, mass = P+N, mass − # P = Neutro

↑ Periods  Valence energy shells →
← INCREASING REACTIVITY →

| COMPOUND | FORMULA | ELEMENTS |
|---|---|---|
| Carbon dioxide | $CO_2$ | Carbon, Oxygen |
| Water | $H_2O$ | Hydrogen, Oxygen |
| Hydrogen Sulfide | $H_2S$ | Hydrogen, Sulfur |
| Methane | $CH_4$ | Carbon, Hydrogen |
| Sodium Chloride | $NaCl$ | Sodium, Chlorine |
| Hydrogen Peroxide | $H_2O_2$ | Hydrogen, Oxygen |
| Hydrochloric acid | $HCl$ | Hydrogen, Chlorine |
| Carbon Monoxide | $CO$ | Carbon, Oxygen |

**most abundant elements CHONPS**

More than one different kind of element chemically combined is a compound.

$C_6H_{12}O_6$ (sugar)

Subscripts identify the number of atoms of each element.

LiF

## FAST FORMULAS

- Force: F = M · A
- Density: M = D · V
- Speed: D = S · T
- Work: W = F · D

## Moon Phases:
Full Moon, Waxing Crescent, 1st Quarter, Waxing Gibbous, New Moon, Waning Gibbous, 3rd Quarter, Waning Crescent — Earth, SUN

THERMAL ENERGY ALWAYS MOVES FROM WARMER TO COOLER.

## Taxonomic Classification
DOMAIN, KINGDOM, PHYLUM, CLASS, ORDER, FAMILY, GENUS

## ELECTROMAGNETIC SPECTRUM
Low Frequency ← increasing energy / increasing wavelength → High Frequency

Radio, Microwave, Infrared, Visible Light, Ultraviolet, X-Ray, Gamma

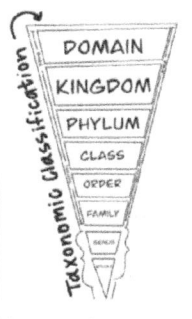

Spunky Science ©

# INTERPRETING DATA

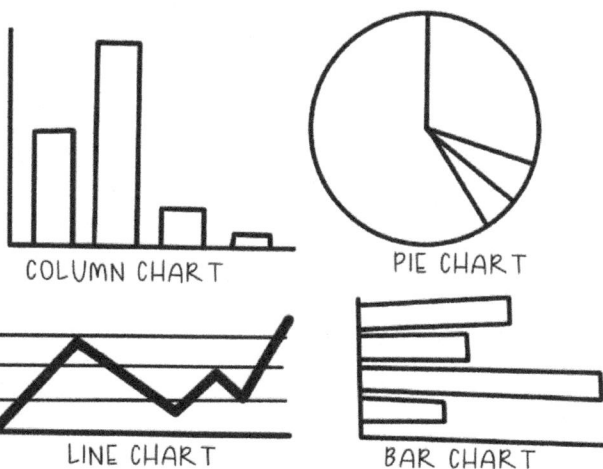

## CHARTS

COLUMN CHART
PIE CHART
LINE CHART
BAR CHART

Charts are tools that organize information in order to easily see trends, make comparisons, and see the meaning behind the numbers.

## TABLES

| Element Abundance | % |
|---|---|
| Oxygen | 47 |
| Silicon | 28 |
| Aluminum | 8 |
| Iron | 5 |
| Calcium | 4 |
| Sodium | 3 |
| Potassium | 2.6 |
| Mgnsium | 2 |

A table is a set of data that is organized using vertical columns and horizontal rows.

## GRAPHS

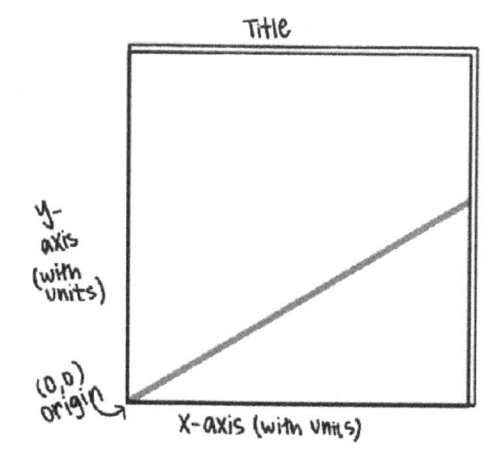

Title
y-axis (with units)
(0,0) origin
X-axis (with units)

Graphs take information, organize it, and present it as a "picture" of your data.

## MAPS

A map is a representation of Earth's features drawn on a flat surface. Symbols and colors represent important features on a map.

**CORRELATION**
A mutual relationship or connection between two or more things.

**CAUSATION**
When one thing or act directly causes another.

- Shark Attacks
- Ice Cream Sales

numbers

months of the year

# ICE CREAM CONSUMPTION IS CAUSING SHARK ATTACKS

# CORRELATION DOES NOT IMPLY CAUSATION

# KINETIC
## ENERGY TYPES

© Spunky science

**Mechanical** — Energy that moves (a machine or moving part)

**Electrical** — Caused by a flow of electrons

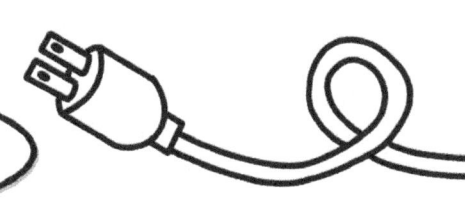

**Light** 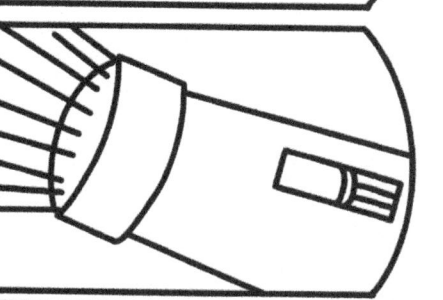 — Travels in waves through space

**Thermal** — Heat created by molecules moving quickly

YOU'RE TOAST

**Sound** — Waves made by vibrating objects

# VISIBLE LIGHT

**Amplitude:** One half of the distance between a waves high point and low point. The more energetic waves have a higher have a larger amplitude.

**Trough:** The lowest point that a wave reaches from its resting point.

**Crest:** The highest point that a wave reaches from its resting point.

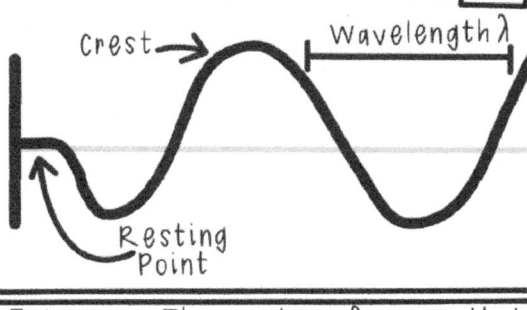

Crest → Wavelength λ  Amplitude  Resting Point  Trough

**Frequency:** The number of waves that pass a fixed point in a given unit of time. Written as f with units in hz.

**Wavelength:** Either measured from crest to crest or trough to trough and written as λ.

Wave Speed = frequency × wavelength
$V = f \times \lambda$

An x-ray technician uses their knowledge of the electromagnetic spectrum to take and interpret x-rays taken from a patient.

The human eye can detect λ from 380-700 nm. That's why we can see colors, but not other forms of waves. This part of the ES is called visible light and ranges from 380-740 nm.

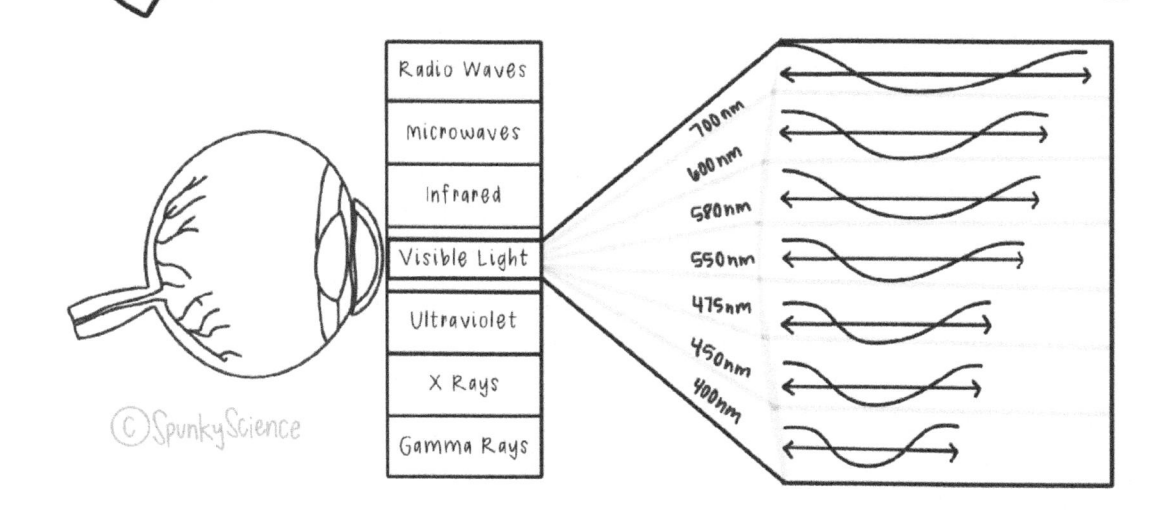

| Radio Waves |
| Microwaves |
| Infrared |
| Visible Light |
| Ultraviolet |
| X Rays |
| Gamma Rays |

700nm
600nm
580nm
550nm
475nm
450nm
400nm

©SpunkyScience

# POTENTIAL ENERGY

PE is stored energy that depends on the relative position of various system parts.

**WITH MASS AND HEIGHT** ↑

**CHEMICAL:** Energy that is available for release from chemical reactions

**ELASTIC:** Occurs when objects are impermanently compressed, stretched or deformed. (Rubber band)

**GRAVITATIONAL:** The energy stored in an object due to its height (waterfall)

**NUCLEAR:** The energy released during nuclear fission or fusion. → 4 He

©Spunky science

# ELASTIC ENERGY

When released, becomes kinetic

Tension on the band increases stored energy

## ELASTIC or SPRING

$E_{elastic} = 1/2 kx^2$

**Examples:**
- AN ARCHERS STRETCHED BOW
- The spring on a wind up clock
- A stretched rubber band

Elastic energy is a form of POTENTIAL energy because it is STORED in the bonds between atoms in an object or substance when it is temporarily under stress

© Spunky science

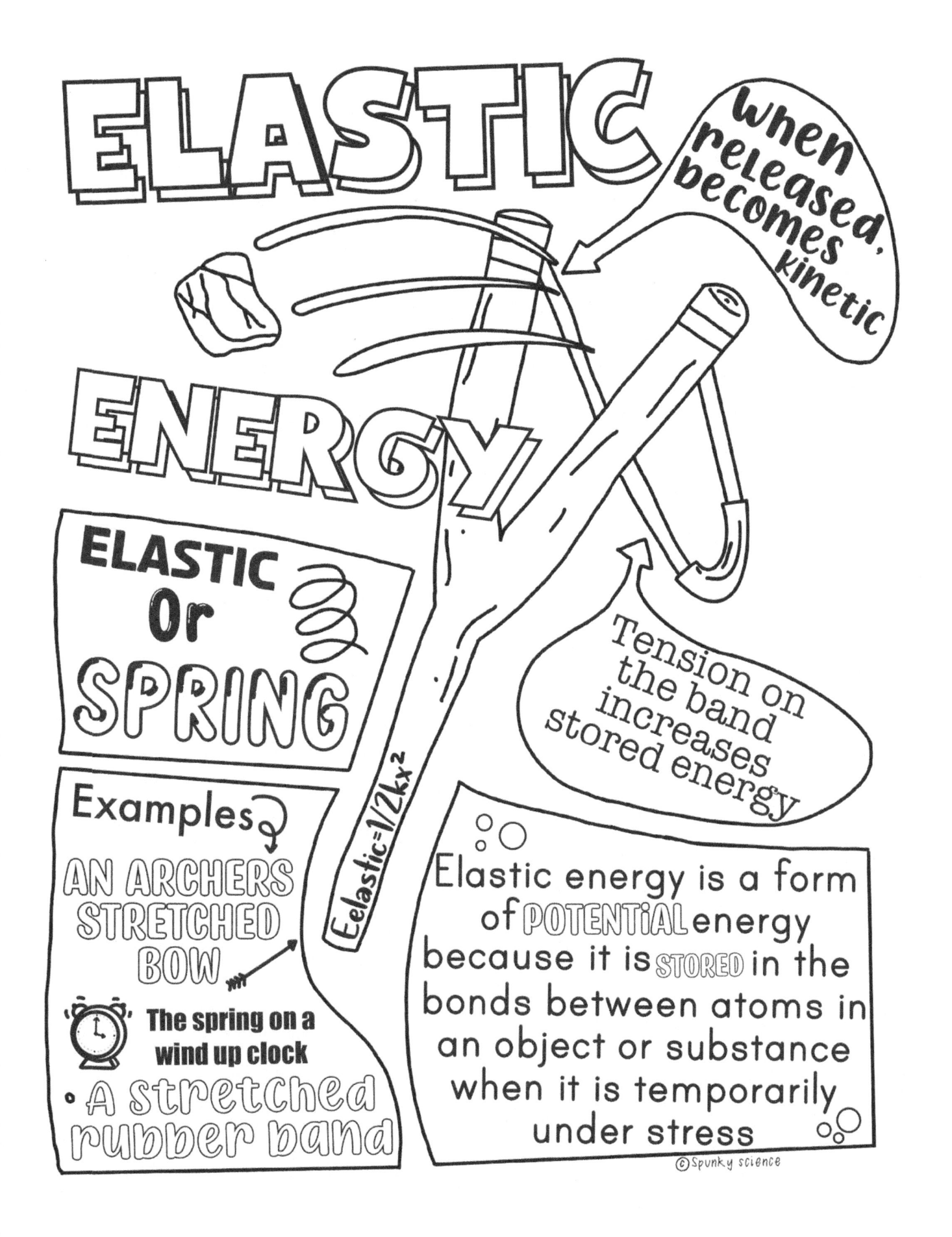

# HOT COCOA HEAT TRANSFER

*SpunkyScience©*

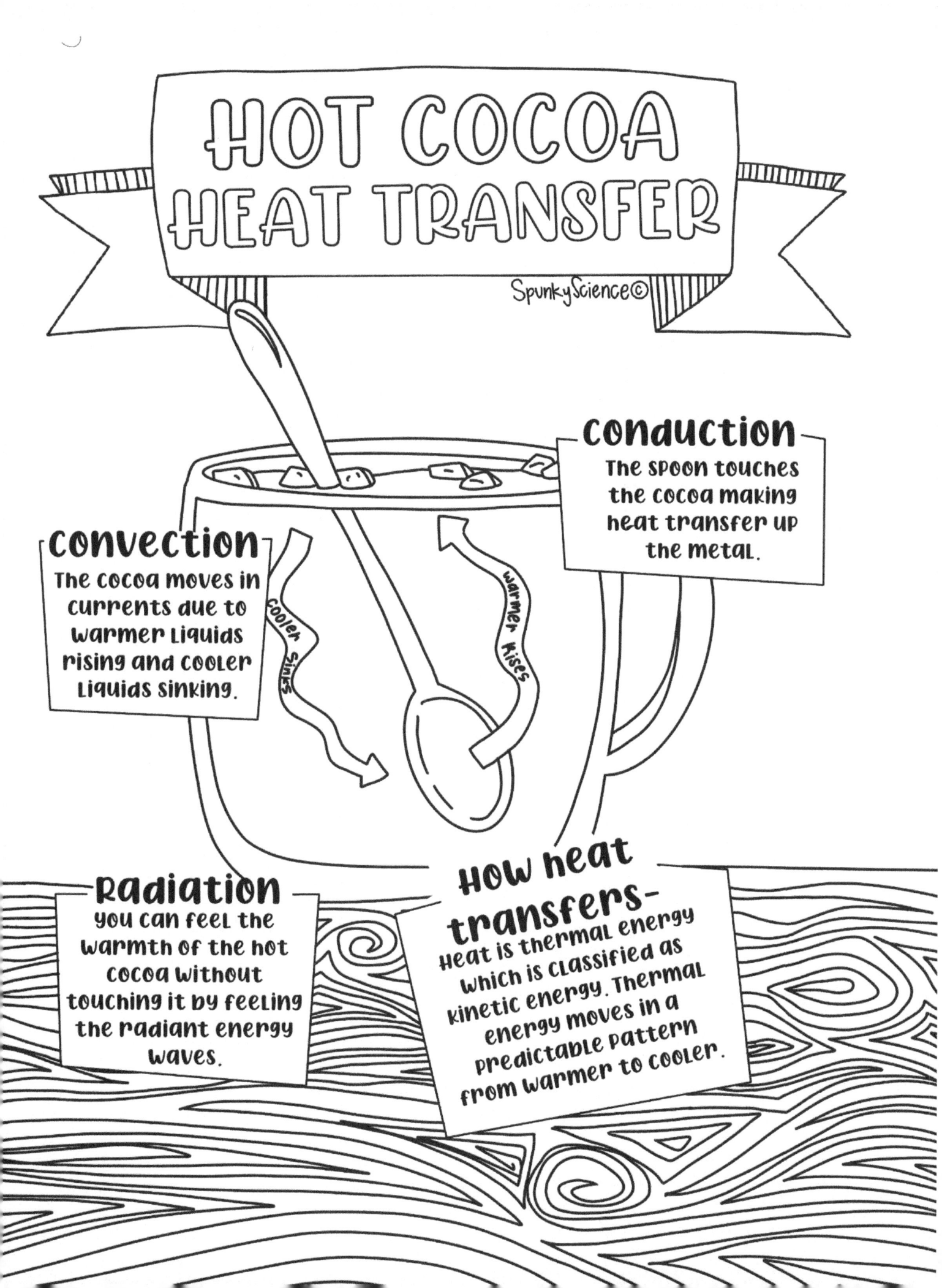

## Conduction
The spoon touches the cocoa making heat transfer up the metal.

## Convection
The cocoa moves in currents due to warmer liquids rising and cooler liquids sinking.

*cooler sinks*
*warmer rises*

## Radiation
You can feel the warmth of the hot cocoa without touching it by feeling the radiant energy waves.

## How heat transfers-
Heat is thermal energy which is classified as kinetic energy. Thermal energy moves in a predictable pattern from warmer to cooler.

# HOW ENERGY TRANSFORMS in a Lightbulb!

4. Heat
3. Light
1. Chemical
2. Electrical

©Spunky science

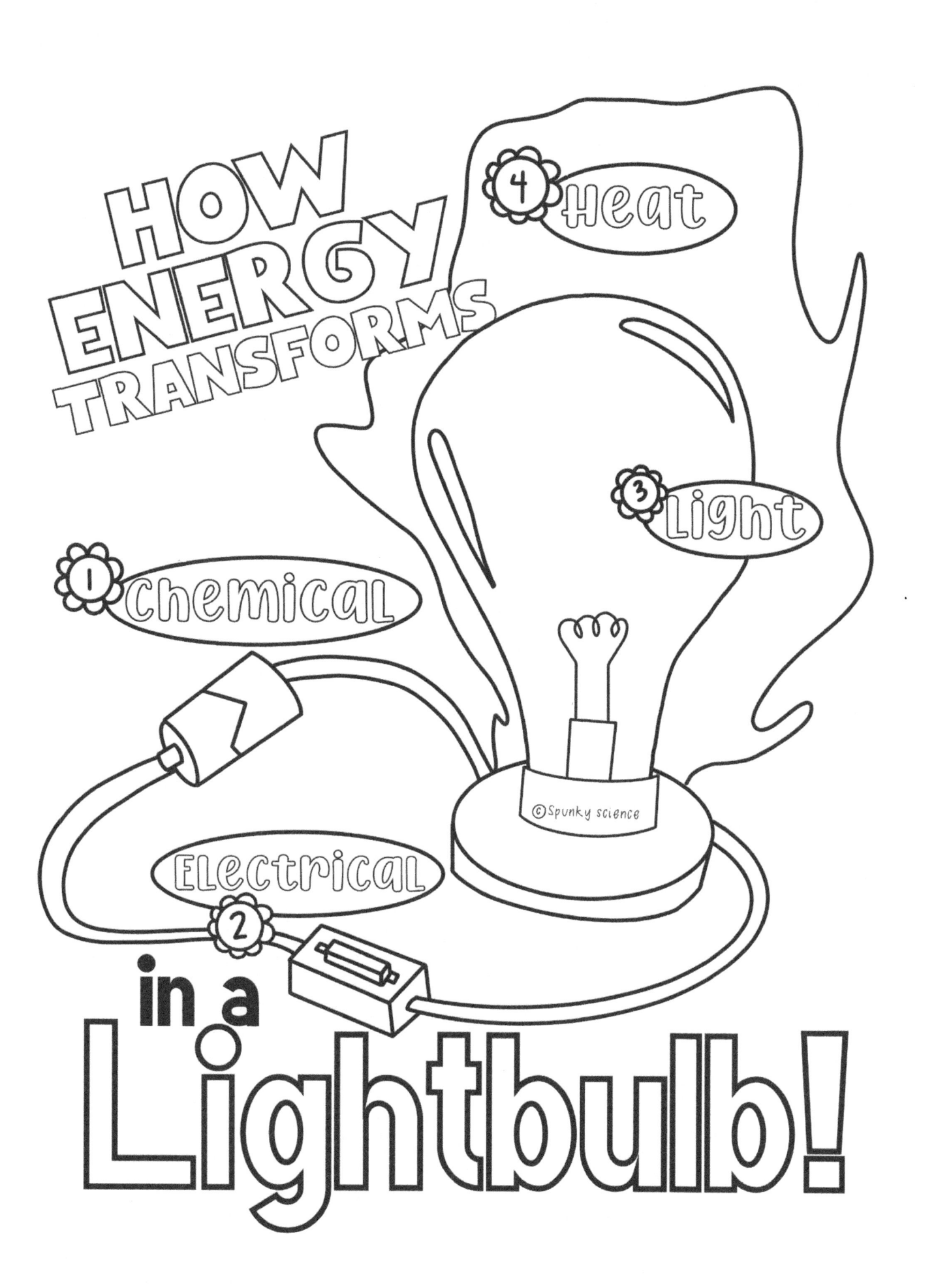

# SNACK MACHINE
## ENERGY TRANSFORMATIONS

**HOW DOES ENERGY TRANSFORM WHEN A SNACK FALLS?**

- MECHANICAL
- LIGHT
- ELECTRICAL
- SOUND

© Spunky science

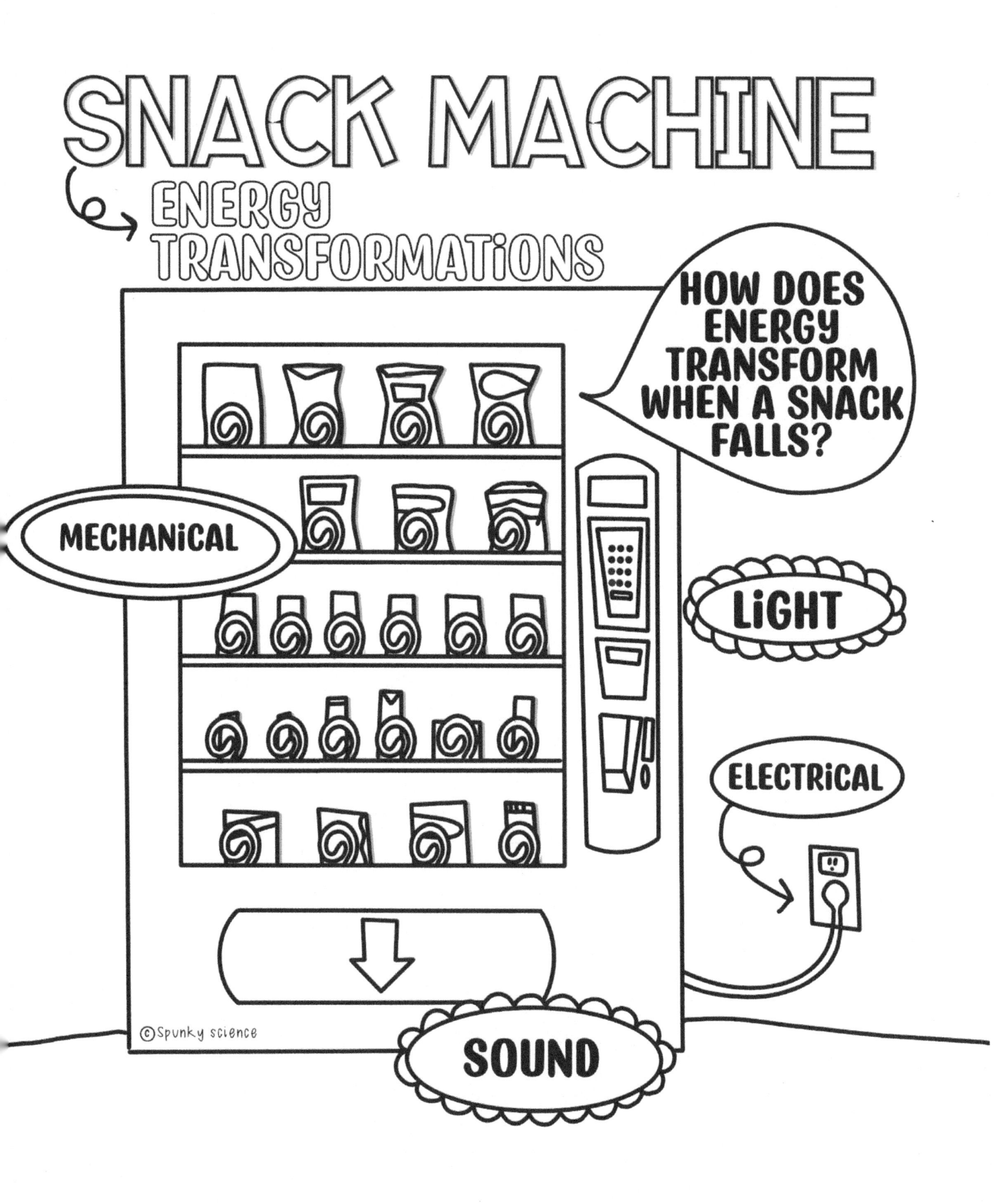

# ENERGY TRANSFORMATIONS

- MECHANICAL
- HEAT
- CHEMICAL
- ELECTRICAL
- RADIANT
- SOUND
- NUCLEAR

CHEMICAL → ELECTRICAL → LIGHT + HEAT

**LAW OF CONSERVATION OF ENERGY**

ENERGY CAN NOT BE CREATED NOR DESTROYED; ENERGY CAN ONLY BE TRANSFERRED OR CHANGED FROM ONE FORM TO ANOTHER.

Spunky Science©

# LAYERS OF THE EARTH

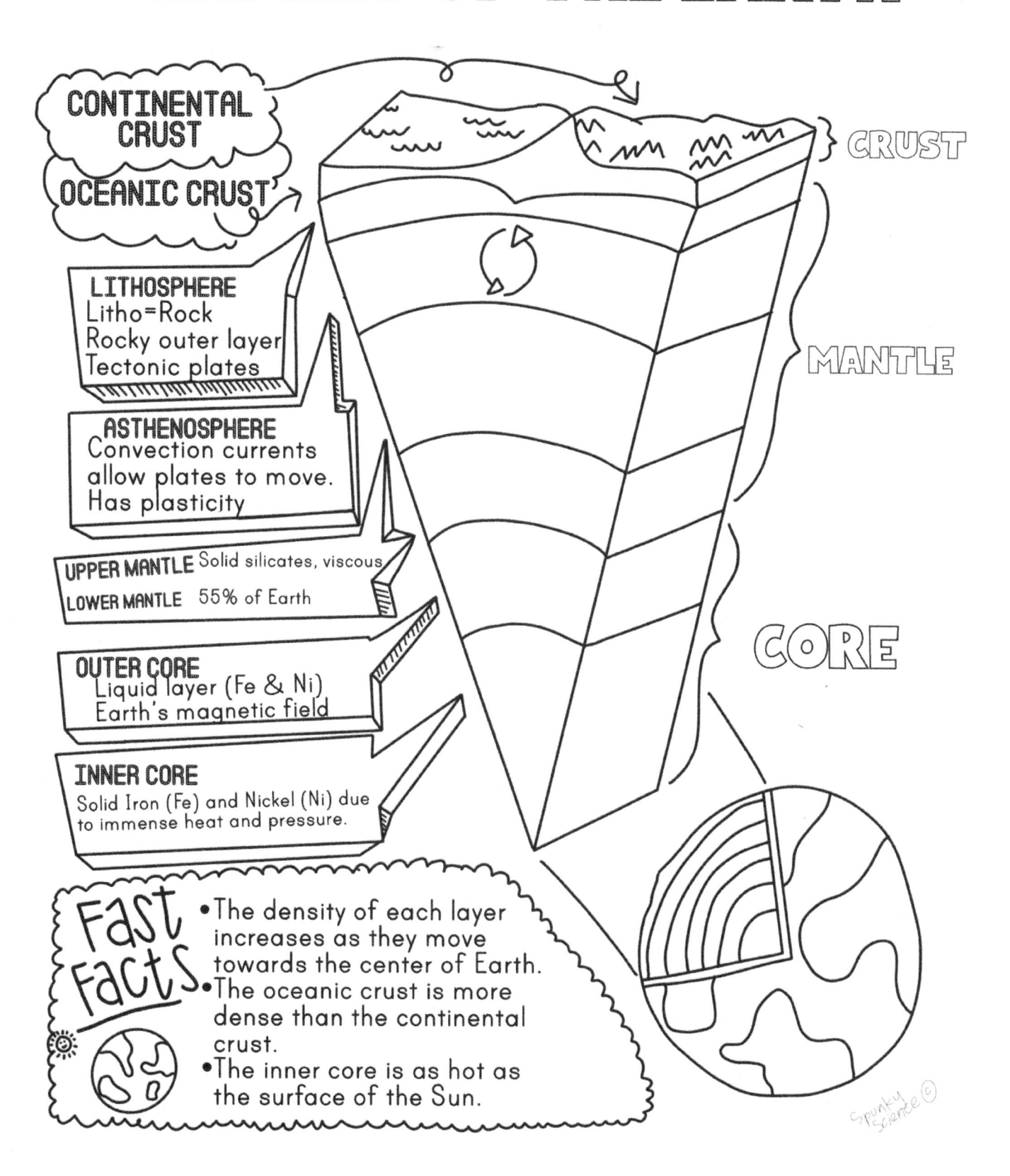

**CONTINENTAL CRUST**

**OCEANIC CRUST**

**LITHOSPHERE**
Litho=Rock
Rocky outer layer
Tectonic plates

**ASTHENOSPHERE**
Convection currents allow plates to move.
Has plasticity

**UPPER MANTLE** Solid silicates, viscous
**LOWER MANTLE** 55% of Earth

**OUTER CORE**
Liquid layer (Fe & Ni)
Earth's magnetic field

**INNER CORE**
Solid Iron (Fe) and Nickel (Ni) due to immense heat and pressure.

CRUST

MANTLE

CORE

**Fast Facts**
- The density of each layer increases as they move towards the center of Earth.
- The oceanic crust is more dense than the continental crust.
- The inner core is as hot as the surface of the Sun.

Spunky Science ©

# PHYSICAL PROPERTIES OF MATTER

NM

**METALS**

**NON METALS**

**METALLOIDS (ALONG STAIRCASE)**

### CONDUCTIVITY
ALLOWS HEAT AND ELECTRICITY TO EASILY PASS

### MAGNETIC
ATTRACTED TO A MAGNET

### BRITTLE
Sulphur
BREAKS EASILY

### DUCTILE
SPOOL OF COPPER WIRE
CAN BE MADE INTO A WIRE

### MALLEABLE
ALUMINUM FOIL
EASILY HAMMERED INTO SHEETS

### LUSTER
REFLECTS LIGHT

SpunkyScience©

Substances are made of different kinds of atoms, which combine with one another in various ways. Atoms form molecules that range in size from 2-2,000 atoms.

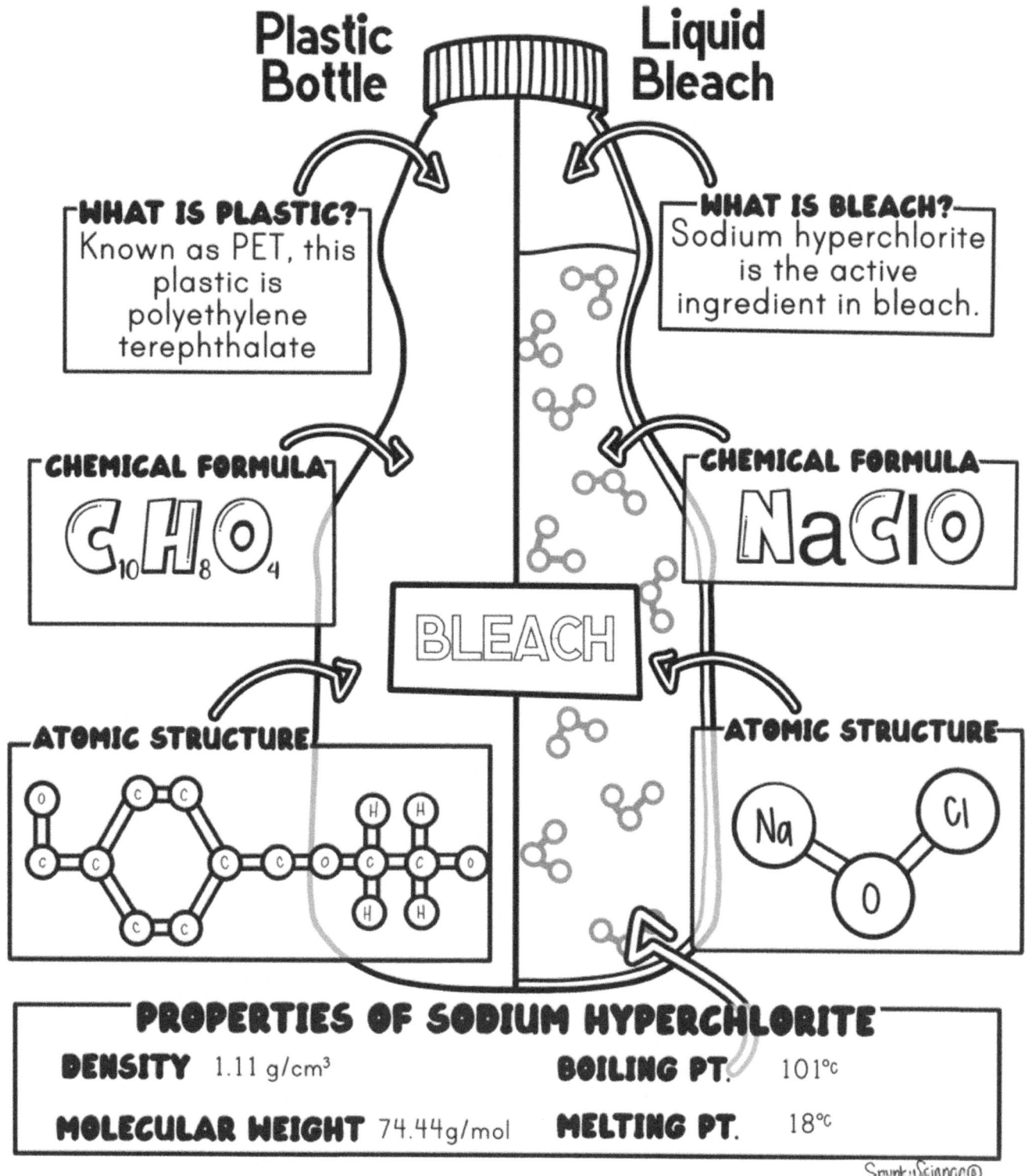

## Plastic Bottle

**WHAT IS PLASTIC?**
Known as PET, this plastic is polyethylene terephthalate

**CHEMICAL FORMULA**
$C_{10}H_8O_4$

**ATOMIC STRUCTURE**

## Liquid Bleach

**WHAT IS BLEACH?**
Sodium hyperchlorite is the active ingredient in bleach.

**CHEMICAL FORMULA**
$NaClO$

**ATOMIC STRUCTURE**

BLEACH

### PROPERTIES OF SODIUM HYPERCHLORITE

**DENSITY** 1.11 g/cm³           **BOILING PT.** 101°c

**MOLECULAR WEIGHT** 74.44g/mol    **MELTING PT.** 18°c

Spunky Science©

# COMPOUNDS : COLOR BY SCIENCE

Test your knowledge of compounds by coloring each space the appropriate color.

## ELEMENT
Color: BLUE SHADES

## COMPOUND
Color: GREEN SHADES

© Spunky Science

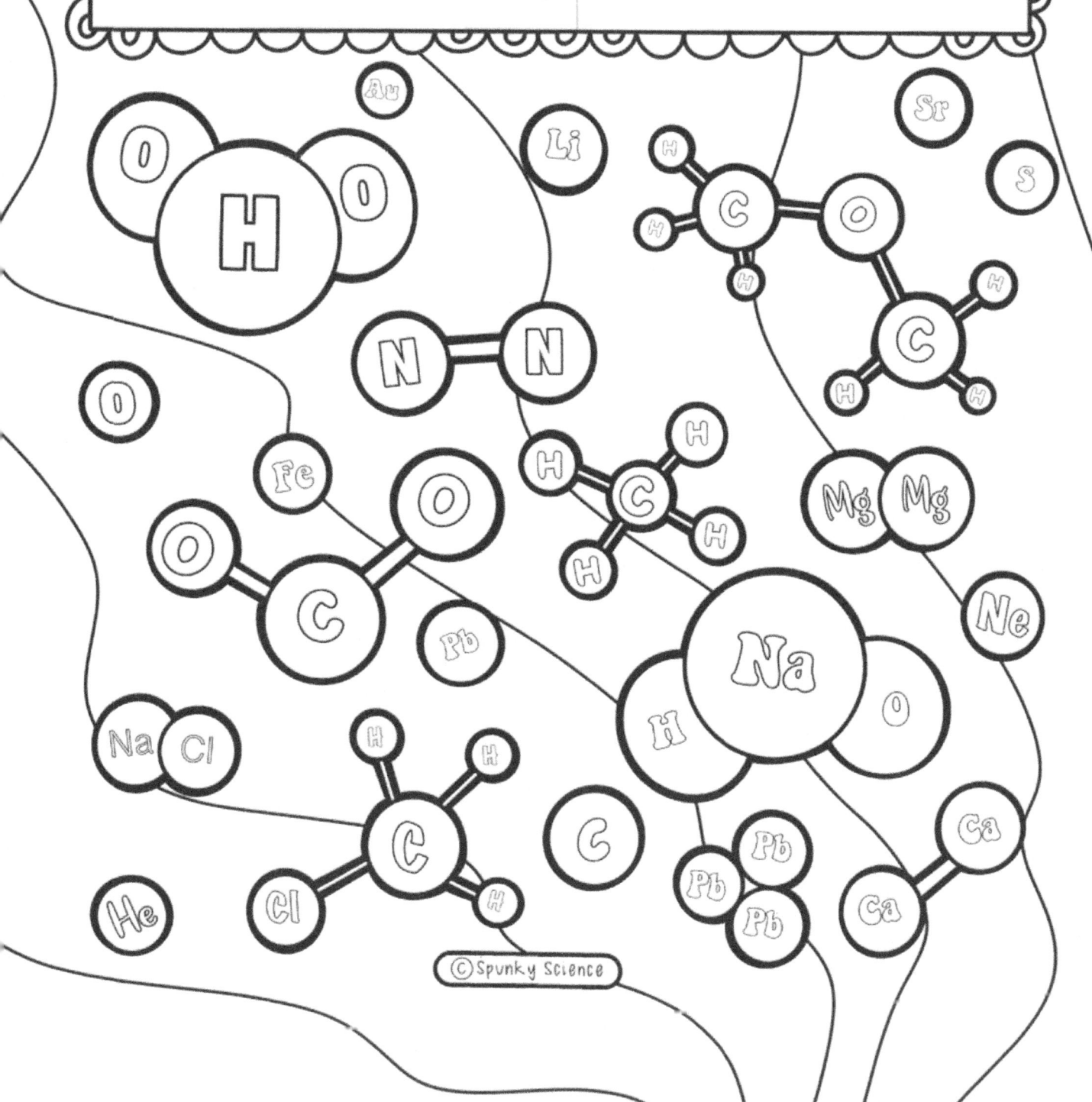

# CHEMICAL REACTIONS

When two or more different substances are mixed, a new substance is formed.

Odor produced

Color change

A Solid is formed

Temperature change

precipitate

Gas created

## VINEGAR AND BAKING SODA ARE A CHEMICAL REACTION

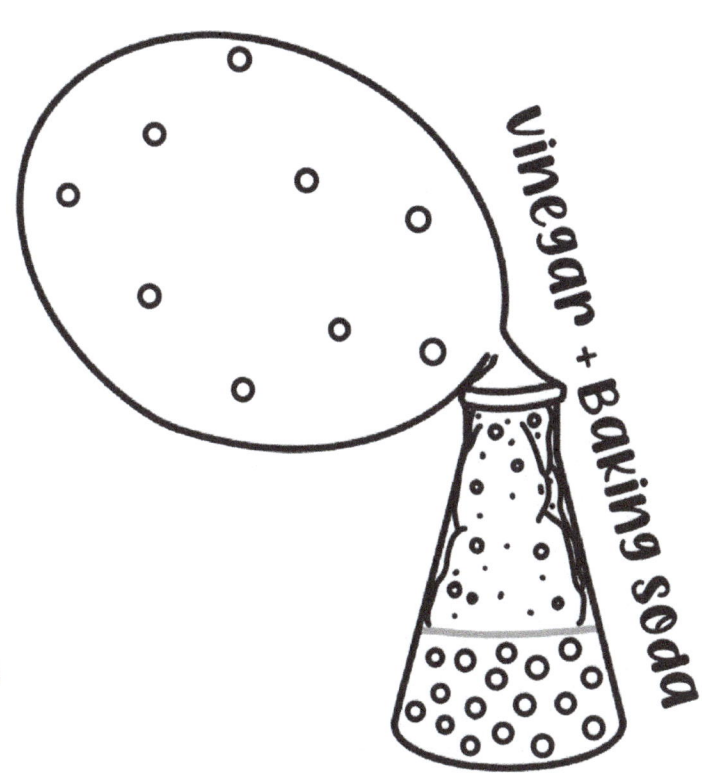

vinegar

vinegar + Baking soda

A new substance is created, but mass is not created nor destroyed.

SpunkyScience©

# CHANGING STATES OF MATTER

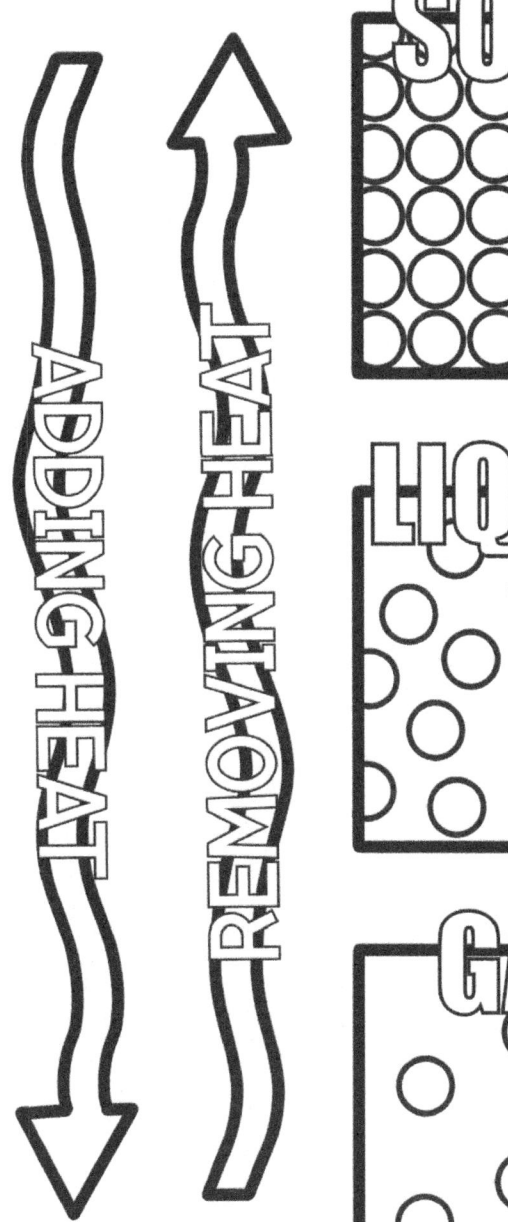

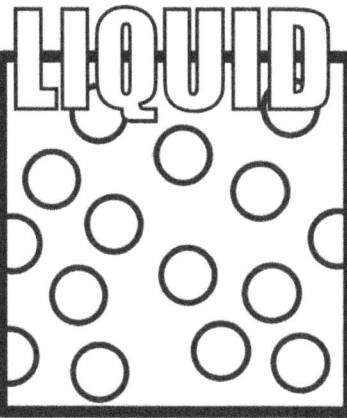

**SOLID** — Tightly packed molecules, vibrate

**LIQUID** — Loosely packed molecules, Takes the shape of the container

**GAS** — Fills space, moves freely

ADDING HEAT ↓

REMOVING HEAT ↑

# STATES OF matter

**Solid**

Freezing · Melting

Sublimation · Deposition

**Liquid**

Evaporation · Condensation

**Gas**

---

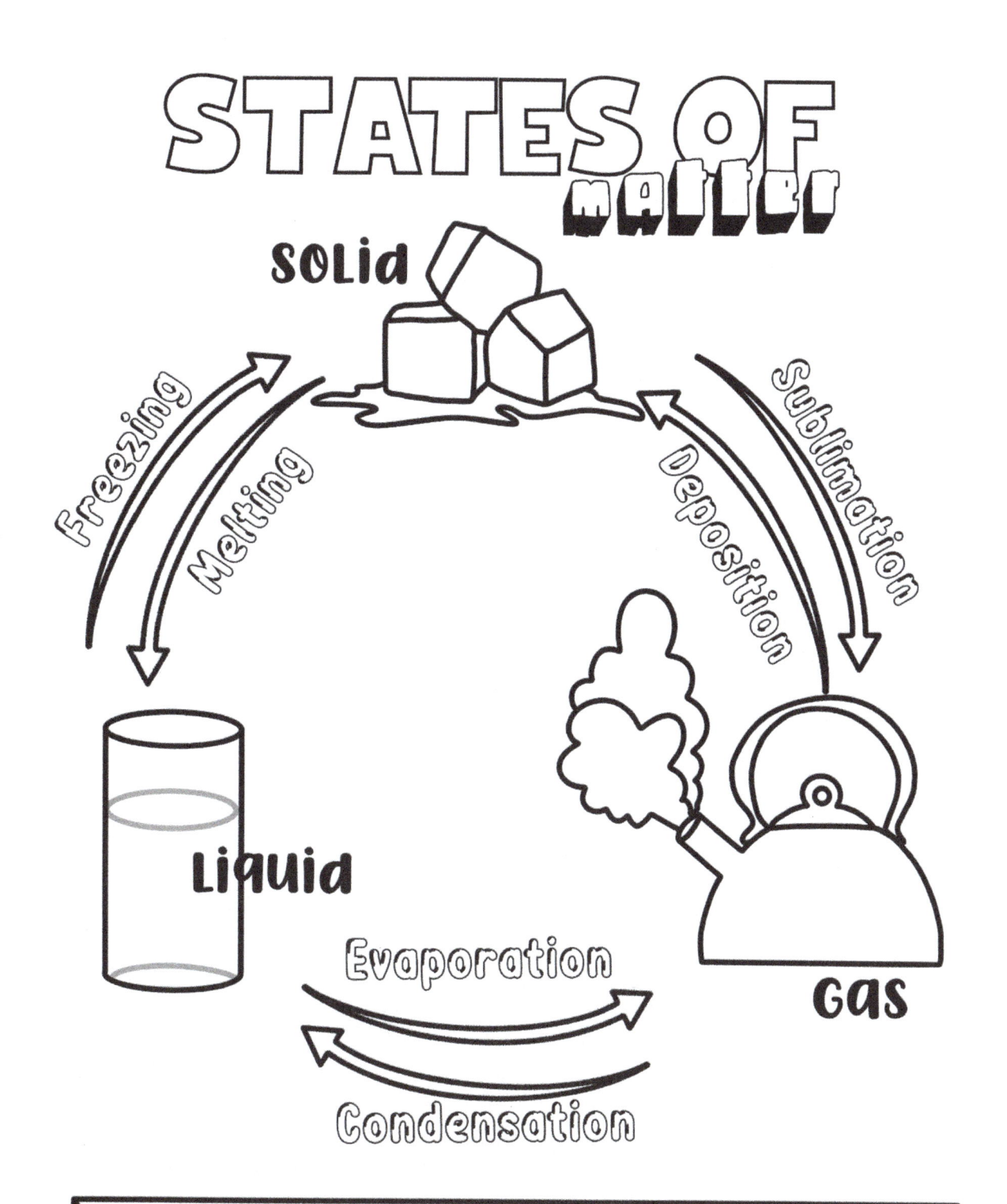

**GASES, LIQUIDS, AND SOLIDS** are all made up of microscopic **PARTICLES,** but the behaviors of these particles are different in each phase.

SpunkyScience©

# UNBALANCED FORCES

NOT ZERO

10N

35N

Causes objects to speed up, slow down, stop, or change direction.

One force is greater than the opposite force

# BALANCED

**Example:** A skateboarder is moving at a constant speed of 12mph

↓ 7N

Object stays still or moves at a CONSTANT SPEED

→ 5N

← 5N

↑ 7N

Two forces that are EQUAL in size and acting in OPPOSITE directions

No ACCELERATION (speeding up)

# FORCES

# ROTATE

SPIN/TURN

AXIS

TAKES 24 HOURS OR ONE DAY

CAUSES DAY AND NIGHT

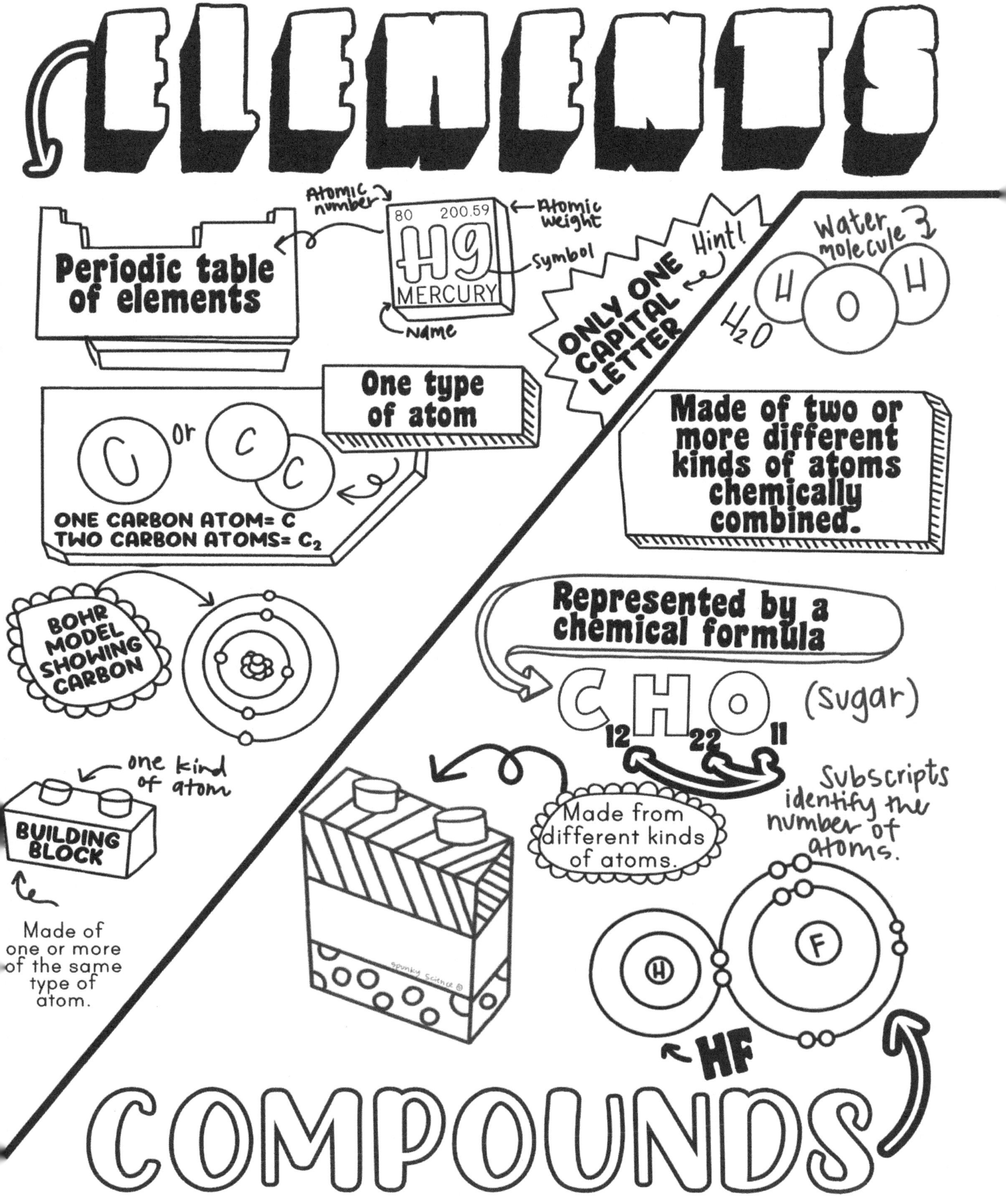

# ELEMENTS

**Periodic table of elements**

Atomic number → 80, 200.59 ← Atomic weight
Hg
MERCURY
Symbol
Name

**Hint! ONLY ONE CAPITAL LETTER**

**One type of atom**

C or C C

ONE CARBON ATOM = C
TWO CARBON ATOMS = $C_2$

**BOHR MODEL SHOWING CARBON**

**BUILDING BLOCK** — one kind of atom

Made of one or more of the same type of atom.

# COMPOUNDS

Water molecule
H O H
$H_2O$

**Made of two or more different kinds of atoms chemically combined.**

**Represented by a chemical formula**

$C_{12}H_{22}O_{11}$ (sugar)

Made from different kinds of atoms.

Subscripts identify the number of atoms.

HF (H — F Bohr model)

# MATTER

Matter can be divided into particles that can't be seen.

The Periodic Table of Elements

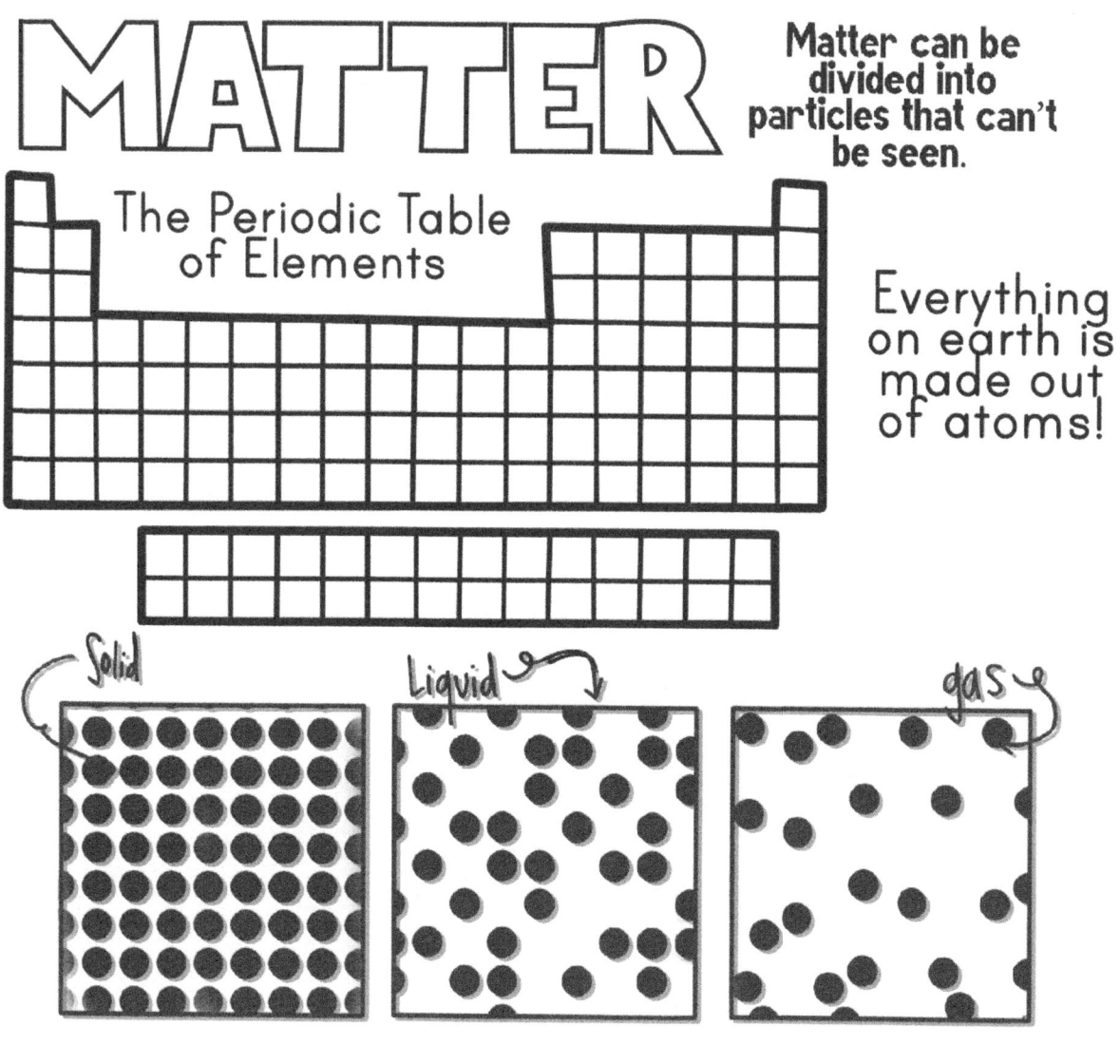

Everything on earth is made out of atoms!

Solid | Liquid | gas

# CONSERVATION OF MATTER

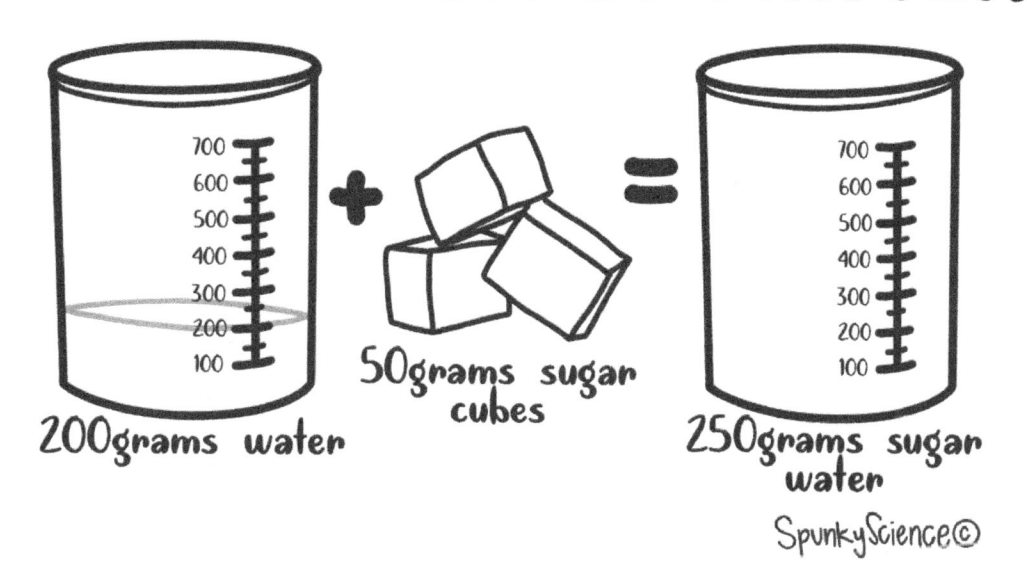

200 grams water + 50 grams sugar cubes = 250 grams sugar water

SpunkyScience©

# ELECTROLYSIS

By adding **electricity** to water and providing a path for the different particles to follow, the water can be **separated** into **HYDROGEN & OXYGEN.**

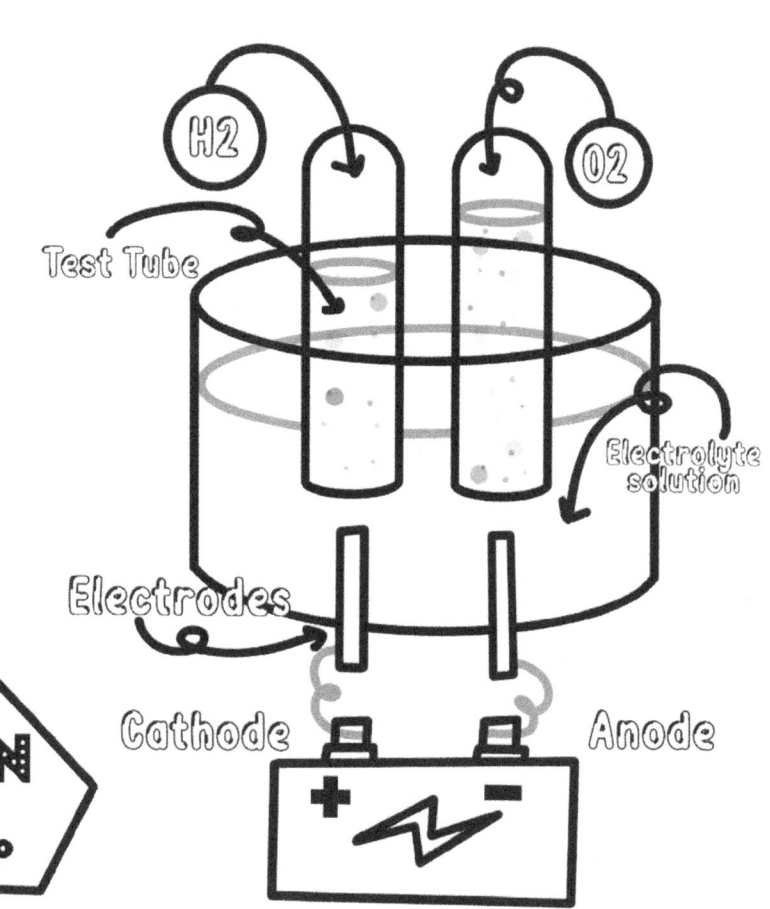

Test Tube
H2
O2
Electrolyte solution
Electrodes
Cathode
Anode
Power Source

## FORMULA

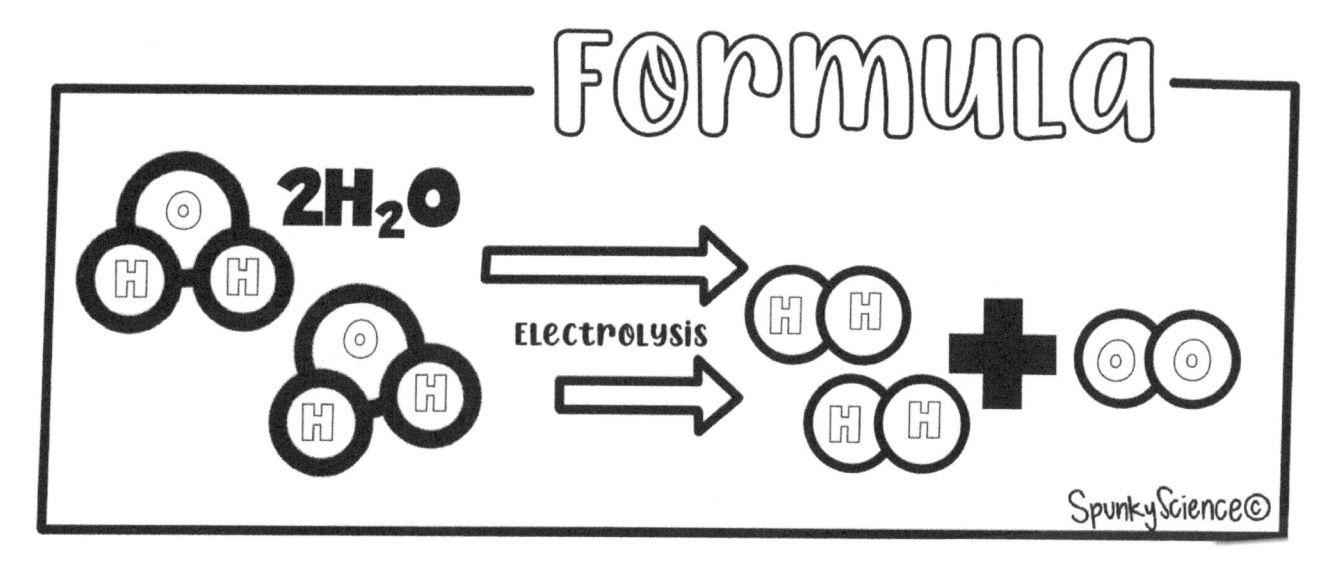

$2H_2O$ → ELECTROLYSIS → H H + H H + O O

SpunkyScience©

# All about ELEMENTS

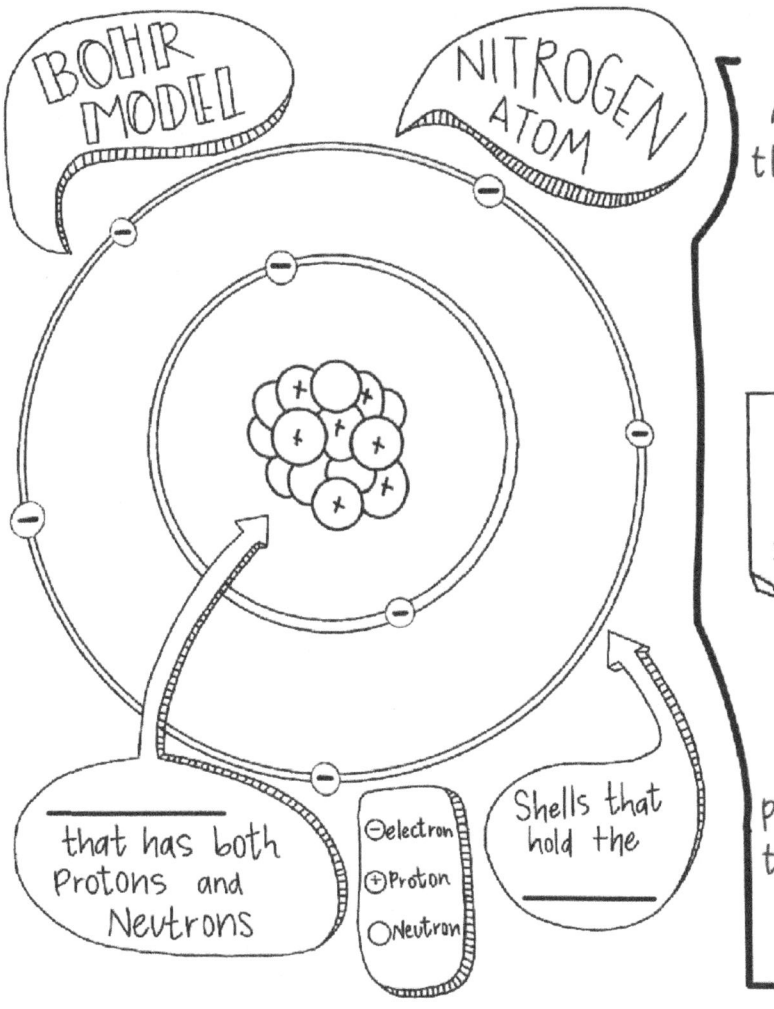

**BOHR MODEL**

**NITROGEN ATOM**

An element is a substance that _____ be broken down into simpler substances. Each atom has a unique number of _____.

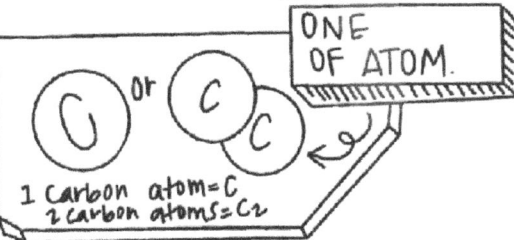

ONE _____ OF ATOM.

1 Carbon atom = C
2 Carbon atoms = C₂

For example, a Nitrogen atom has _____ and a Hydrogen atom has one proton. When more than one type of element is present, it becomes a _____.

_____ that has both Protons and Neutrons

⊖ electron
⊕ proton
◯ neutron

Shells that hold the _____

## Abundance

Out of all of the elements on the periodic table that have been discovered, there are only a few that are most abundance on Earth, in our organisms, and our atmosphere.

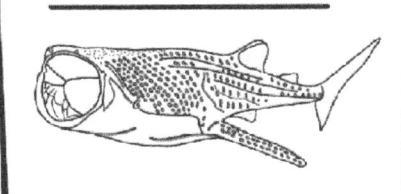

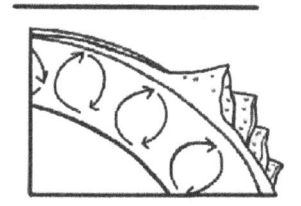

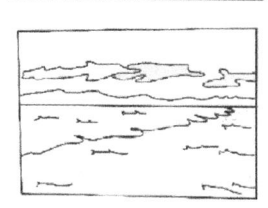

_____   _____   _____   _____

SpunkyScience

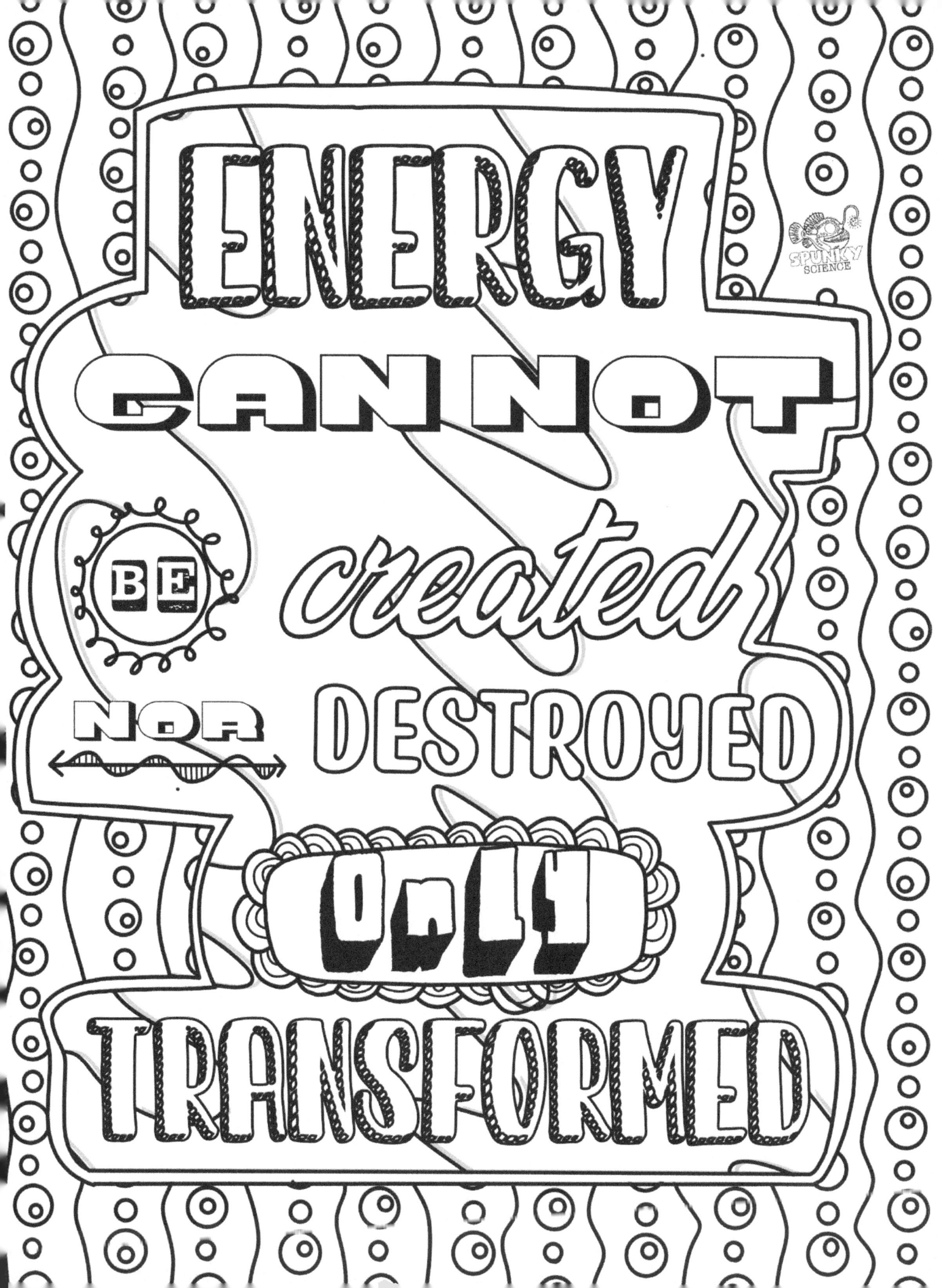

# ENERGY CANNOT BE created NOR DESTROYED ONLY TRANSFORMED

# PERIODIC TABLE

**Atomic Number** → 7
**Symbol** → N
**Name** → Nitrogen
**Atomic Weight** → 14.0067

Each box on the periodic table represents a specific element. Its location on the PT was carefully determined based on each elements physical and chemical properties. These trends make it easy to remember the common properties of elements!

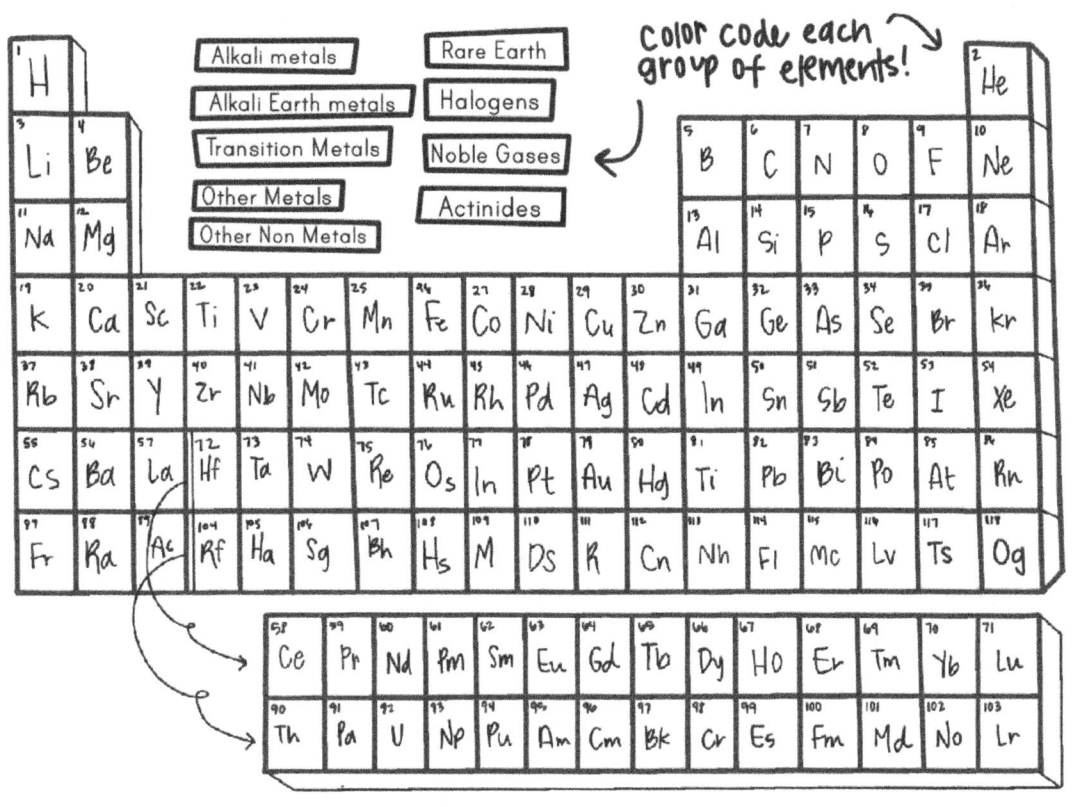

Color code each group of elements!

- Alkali metals
- Alkali Earth metals
- Transition Metals
- Other Metals
- Other Non Metals
- Rare Earth
- Halogens
- Noble Gases
- Actinides

← Color code periods and family →

Each horizontal row is called a <u>period</u> while each vertical column is called a <u>group</u> or <u>family</u>. As you increase the atomic number, you **gain** one proton and one electron.

SpunkyScience©

# ELEMENT MATH

Atomic Number
Symbol
Name
Atomic Weight

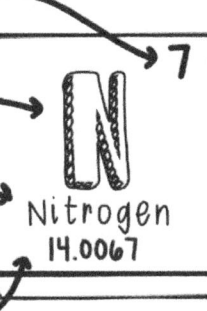

**N** 7
Nitrogen
14.0067

NITROGEN ATOM

Nucleus that has both Protons and Neutrons

⊖ Electron
⊕ Proton
○ Neutron

Shells that hold the electrons

You can easily determine how many protons, neutrons, and electrons that the element has just by looking at the information provided.

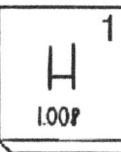

**H** 1
1.008
Hydrogen
Atomic Number: 1
Protons: 1
Electrons: 1
Neutrons: 0

**Li** 3
6.941
Lithium
Atomic Number: 3
Protons: 3
Electrons: 3
Neutrons: 4

**Be** 4
9.012
Berillium
Atomic Number: 4
Protons: 4
Electrons: 4
Neutrons: 5

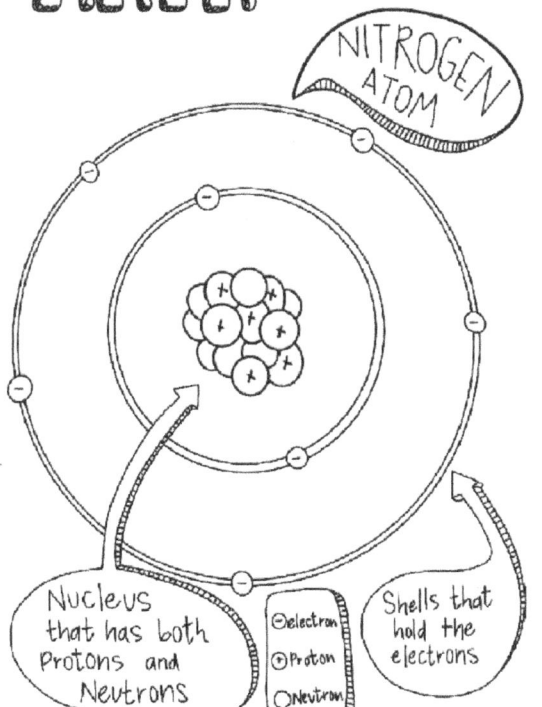

In a neutral atom!

**A**tomic number
EQUALS
**P**rotons
EQUALS
**E**lectrons

The number of protons and electrons are equal to the atomic number.

**M**ass
MINUS
**A**tomic Number
EQUALS
**N**eutrons

The number of neutrons is equal to the atomic number minus the atomic mass.

© SpunkyScience

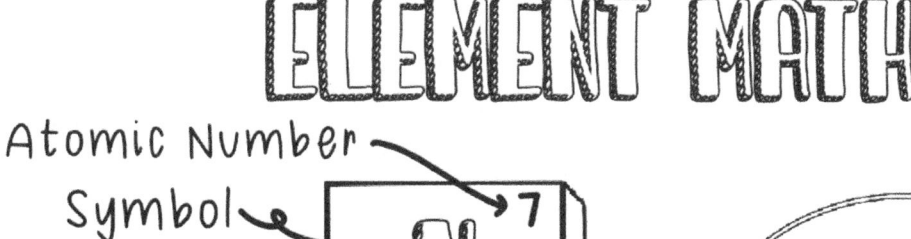

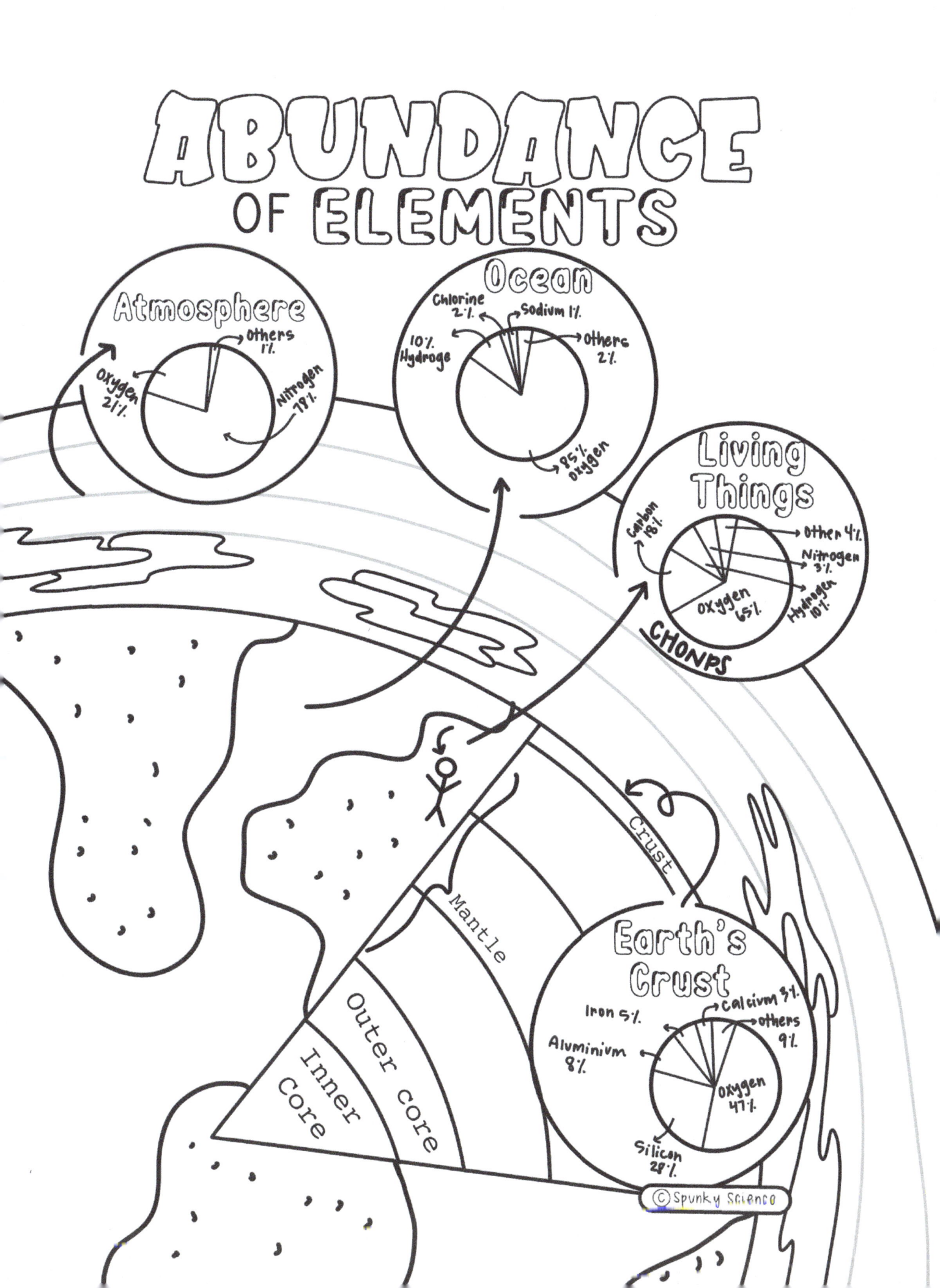

# ABUNDANCE OF ELEMENTS

## Atmosphere
- Oxygen 21%
- Nitrogen 78%
- Others 1%

## Ocean
- Chlorine 2%
- Sodium 1%
- Hydrogen 10%
- Oxygen 85%
- Others 2%

## Living Things
- Carbon 18%
- Oxygen 65%
- Hydrogen 10%
- Nitrogen 3%
- Other 4%

CHONPS

## Earth's Crust
- Iron 5%
- Aluminium 8%
- Calcium 3%
- Others 9%
- Oxygen 47%
- Silicon 28%

Crust
Mantle
Outer core
Inner Core

© Spunky Science

# IDENTIFYING COMPOUNDS

$HC_2H_3O_2$ + $NaHCO$ → $CO_2$ + $H_2O$ + $NaC_2H_3O_2$
(vinegar)   (baking soda)    (water)     (sodium acetate)

### CHEMICAL FORMULA
Uses chemical symbols and numbers to show what a particular compound is made of

### MOLECULES
A group of atoms bonded together of the same element.

### ATOMS
The basic unit of a chemical element

### REACTANTS
Substances at the start of a reaction.

### PRODUCTS
Substances at the end of a reaction.

# MASS CAN NOT BE CREATED NOR DESTROYED ONLY TRANSFORMED

**LAW OF CONSERVATION OF MASS**

© Spunky Science

# AQUEOUS SOLUTIONS

**AN AQUEOUS SOLUTION IS A SOLUTION IN WHICH WATER IS THE SOLVENT**

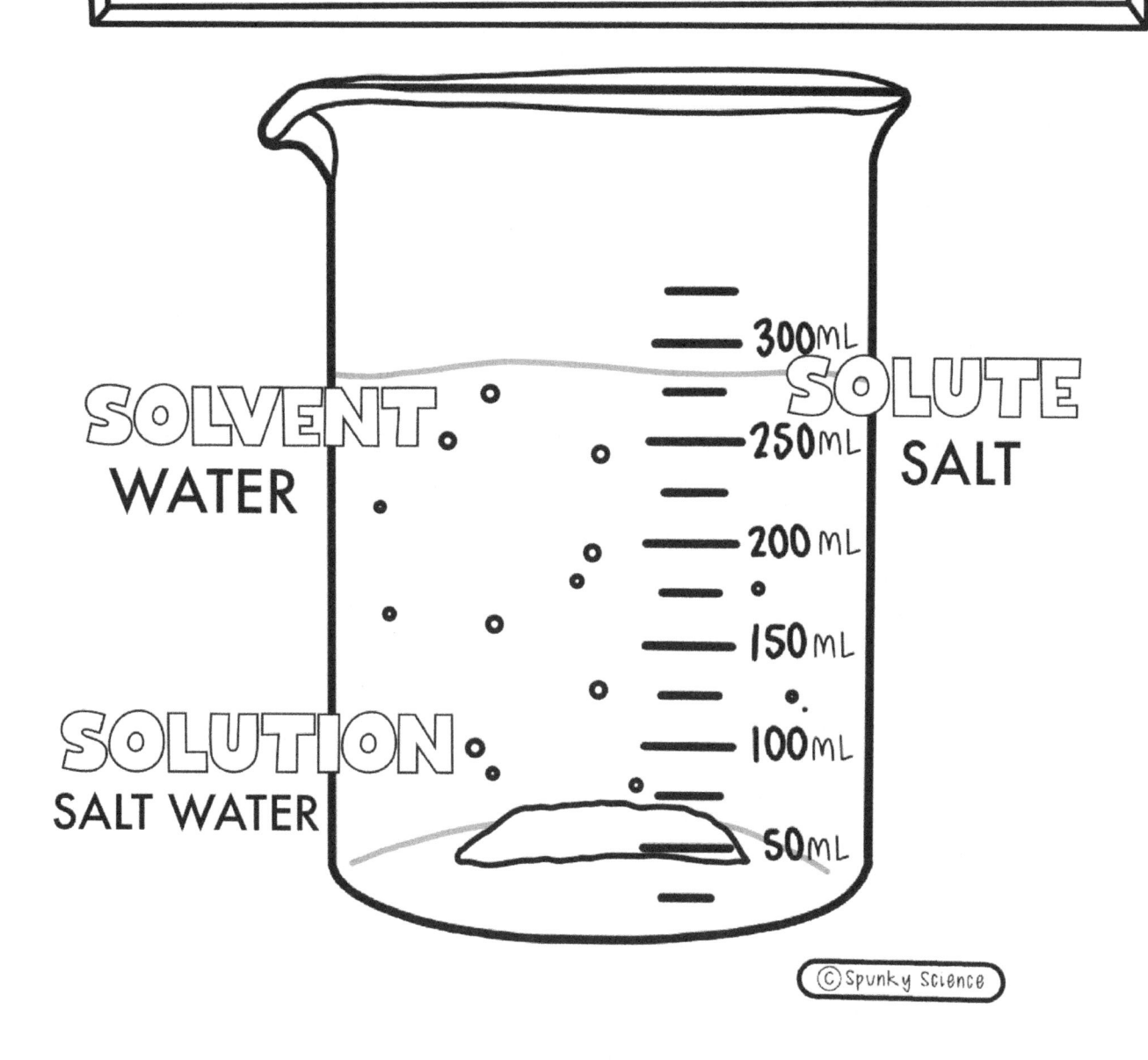

**SOLVENT**
WATER

**SOLUTE**
SALT

**SOLUTION**
SALT WATER

300 mL
250 mL
200 mL
150 mL
100 mL
50 mL

© Spunky Science

# FORCE & MOTION
## ~calculating speed

**FORMULA**

$$S = \frac{D}{T}$$

**UNITS**

**DISTANCE:** *How Far*
- Kilometers
- Miles
- Meters

**TIME:** *How Long*
- Hours
- Minutes
- Seconds

## Average Speed

$$\text{AVERAGE SPEED} = \frac{\text{TOTAL DISTANCE}}{\text{TOTAL TIME}}$$

OR

$$S = \frac{d_2 - d_1}{t_2 - t_1}$$

© Spunky Science

# SPEED CALCULATIONS

## Given
Identify the given information

A cyclist covered a distance of (18 miles) in (45) (minutes). How many miless can he cover in one hour.

## Unknown
What are you trying to solve for?

___?___ miles in one hour

## Equation
What formula do you need to solve the problem?

$$Speed = \frac{distance}{time}$$

## Substitute
Put the numbers that you know into the equation.

$1\,hr = 60\,min$ ← find common units

$$S = \frac{18\,mi}{45\,min} \times \frac{60\,min}{1\,hr} = ?$$

## Solution
Plug the info into the calculator and add your units!

#1  $S = \frac{18\,mi}{45\,min} \times \frac{60\,min}{1\,hr} \times \frac{1{,}080\,mi}{45\,hr}$

#2  multiply tops and bottoms

#3  $\frac{1{,}080\,mi}{45\,hr} = \boxed{24\,mi/hr}$

© Spunky Science

# POSITION-TIME GRAPHS
### * DISTANCE VERSUS TIME *

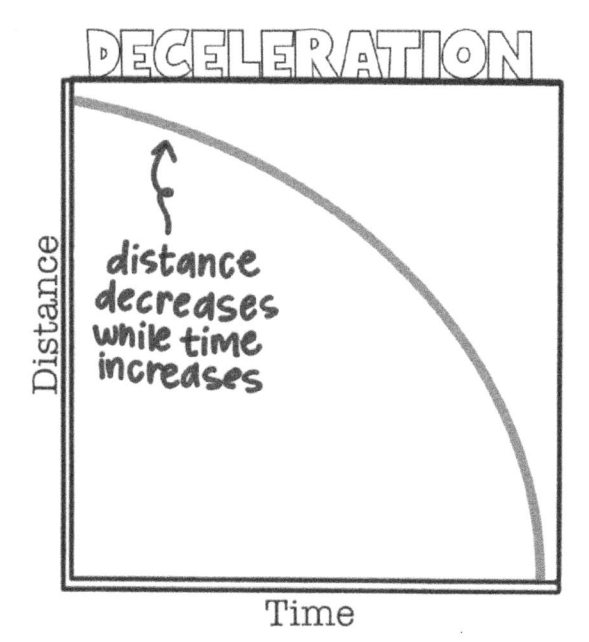

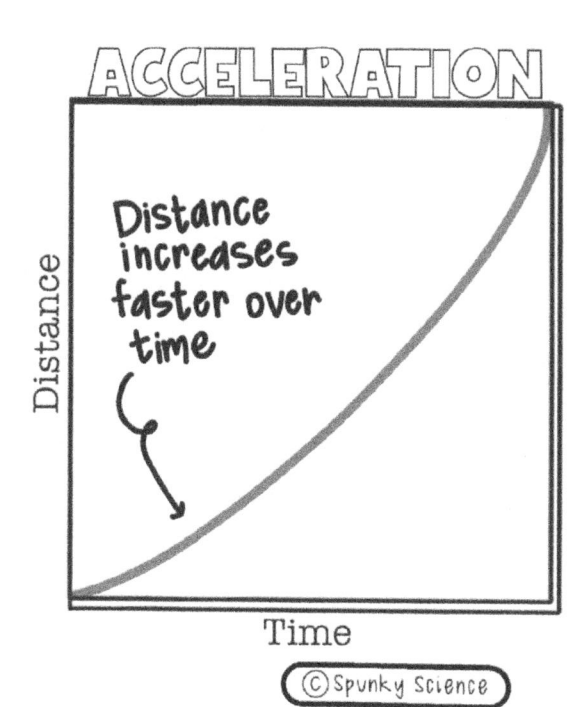

## STATIONARY

(Distance vs. Time graph showing a horizontal line)

distance stays the same while time increases

## CONSTANT SPEED

(Distance vs. Time graph showing a straight diagonal line)

Distance increases at the same rate as time

## DECELERATION

(Distance vs. Time graph showing a curve that flattens out)

distance decreases while time increases

## ACCELERATION

(Distance vs. Time graph showing an upward curving line)

Distance increases faster over time

© Spunky Science

# VELOCITY - TIME GRAPHS

velocity versus time graph shows the speed and direction an object travels over a specific period of time.

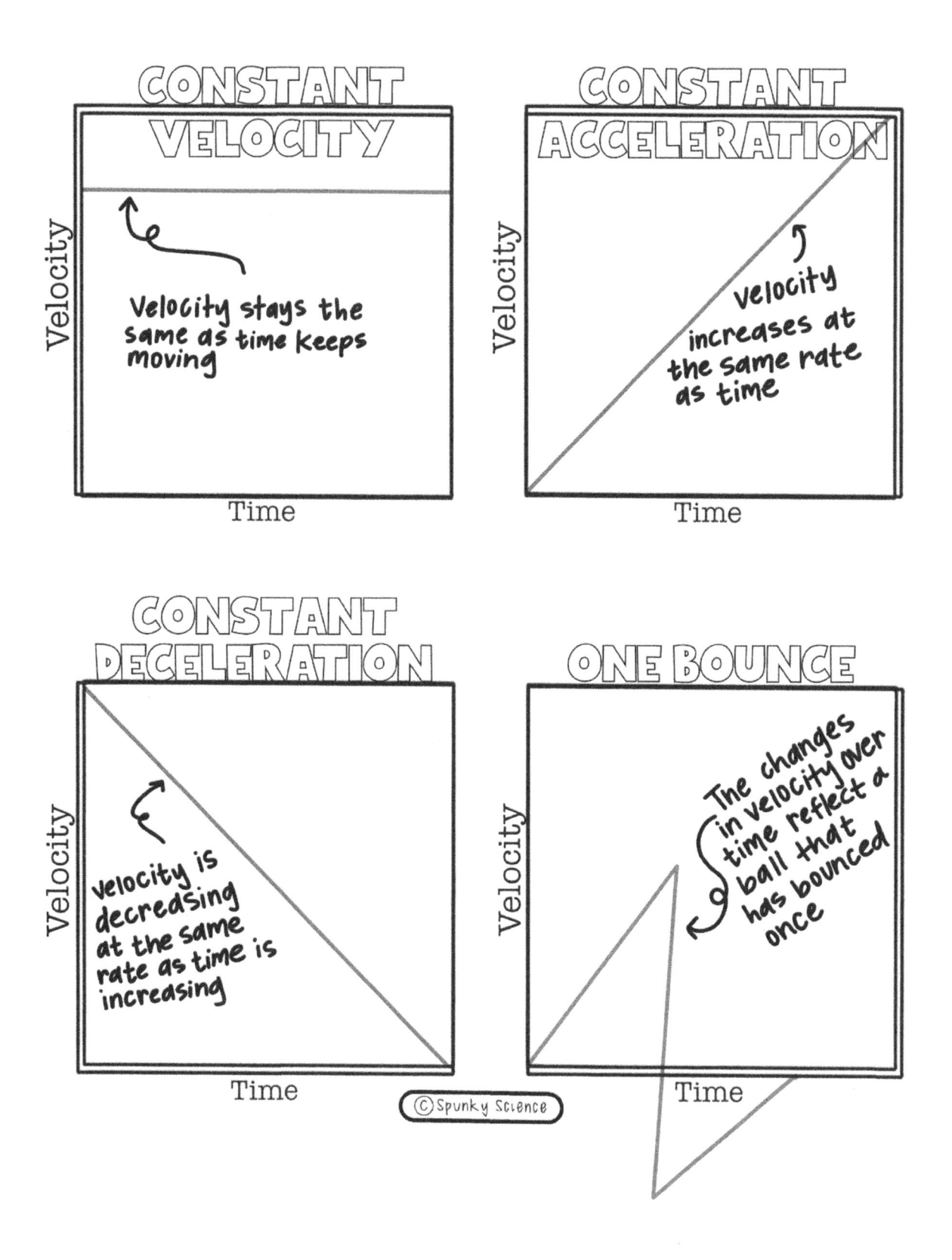

## CONSTANT VELOCITY

Velocity (y-axis) vs Time (x-axis)

Velocity stays the same as time keeps moving

## CONSTANT ACCELERATION

Velocity (y-axis) vs Time (x-axis)

velocity increases at the same rate as time

## CONSTANT DECELERATION

Velocity (y-axis) vs Time (x-axis)

velocity is decreasing at the same rate as time is increasing

## ONE BOUNCE

Velocity (y-axis) vs Time (x-axis)

The changes in velocity over time reflect a ball that has bounced once

© Spunky Science

# VELOCITY

**Velocity is the rate of motion in a specific direction.**

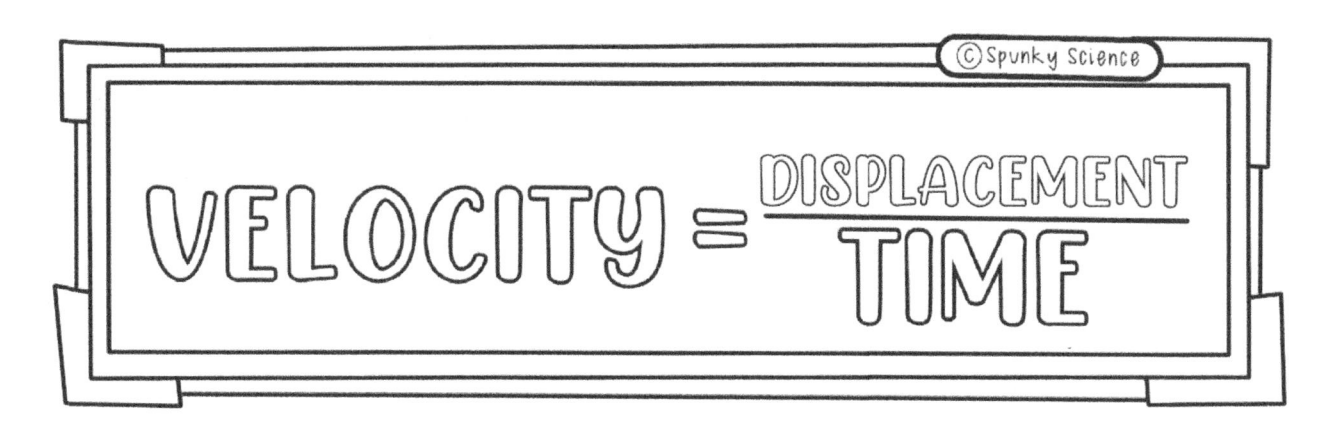

$$\text{VELOCITY} = \frac{\text{DISPLACEMENT}}{\text{TIME}}$$

©Spunky Science

The four different types of velocities are uniform velocity, variable velocity, average velocity, and instantaneous velocity.

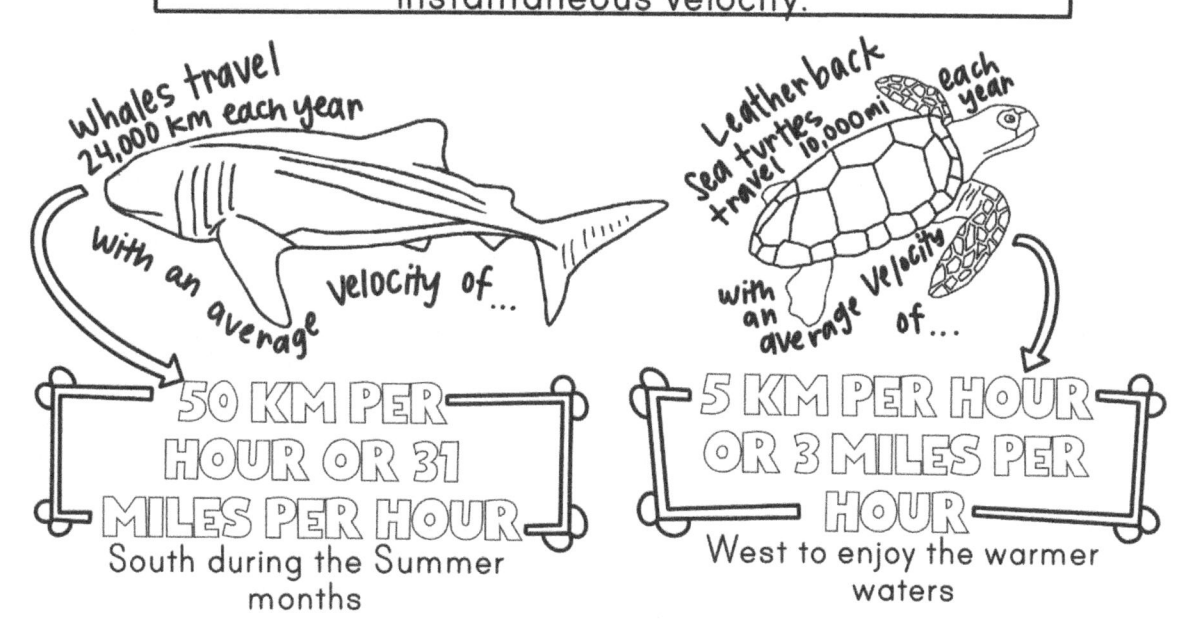

Whales travel 24,000 km each year with an average velocity of...

**50 KM PER HOUR OR 31 MILES PER HOUR**

South during the Summer months

Leatherback sea turtles travel 10,000mi each year with an average velocity of...

**5 KM PER HOUR OR 3 MILES PER HOUR**

West to enjoy the warmer waters

# NEWTON'S FIRST LAW

**AN OBJECT IN MOTION WILL STAY IN MOTION UNLESS ACTED UPON BY AN UNBALANCED FORCE**

An astronaut who lets go of a tool in space will notice that the tool will float in the same speed and direction forever due to the lack of friction and gravity.

A bike that is pushed encounters friction, gravity, and is ultimately stopped by a rock.

# NEWTONS SECOND LAW OF MOTION

The acceleration of an object depends on the objects mass and magnitude or the force acting upon it.

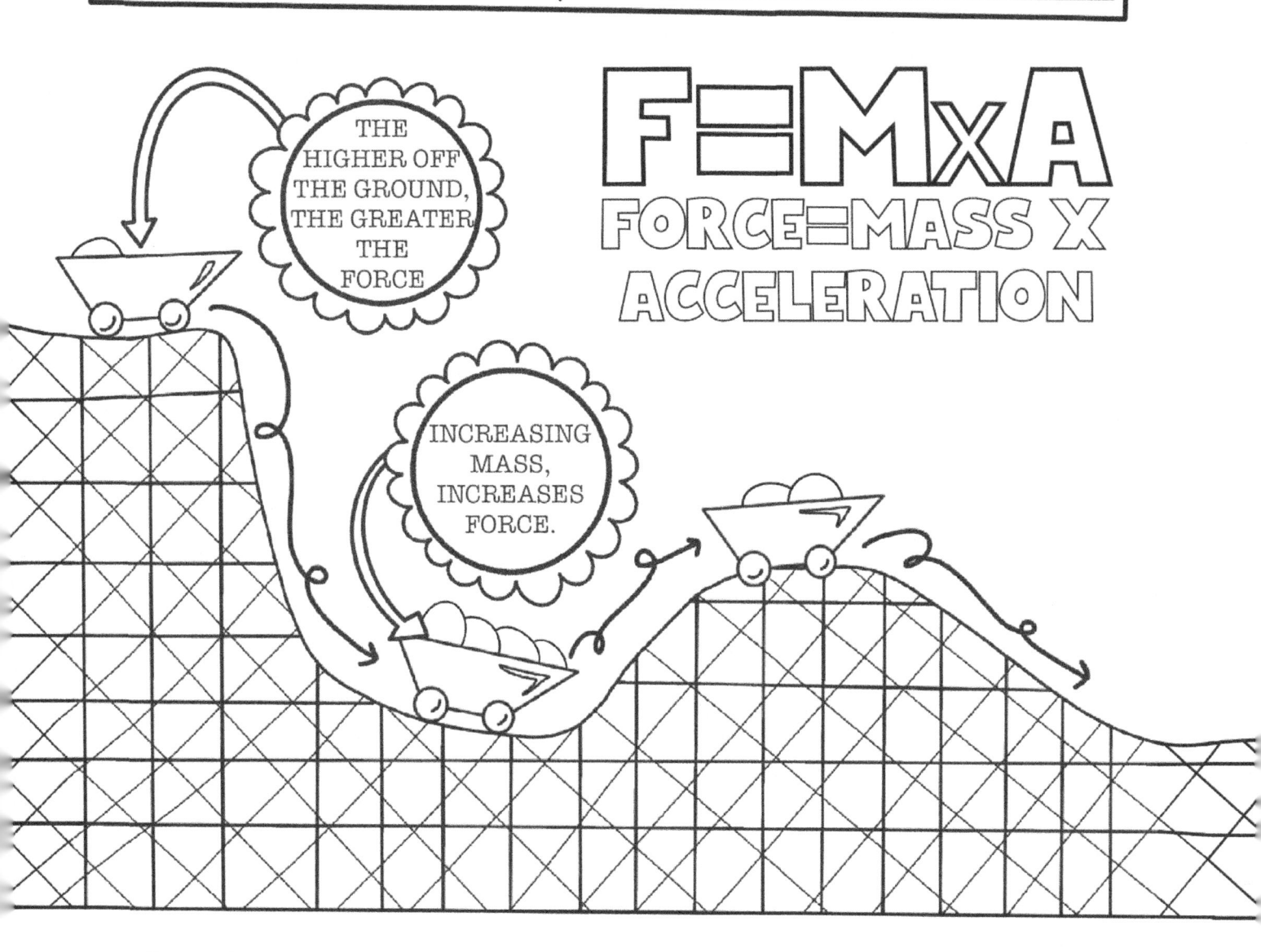

THE HIGHER OFF THE GROUND, THE GREATER THE FORCE

INCREASING MASS, INCREASES FORCE.

# F=MxA
## FORCE=MASS X ACCELERATION

© Spunky Science

# NEWTONS THIRD LAW OF MOTION

For every action there is an equal and opposite reaction.

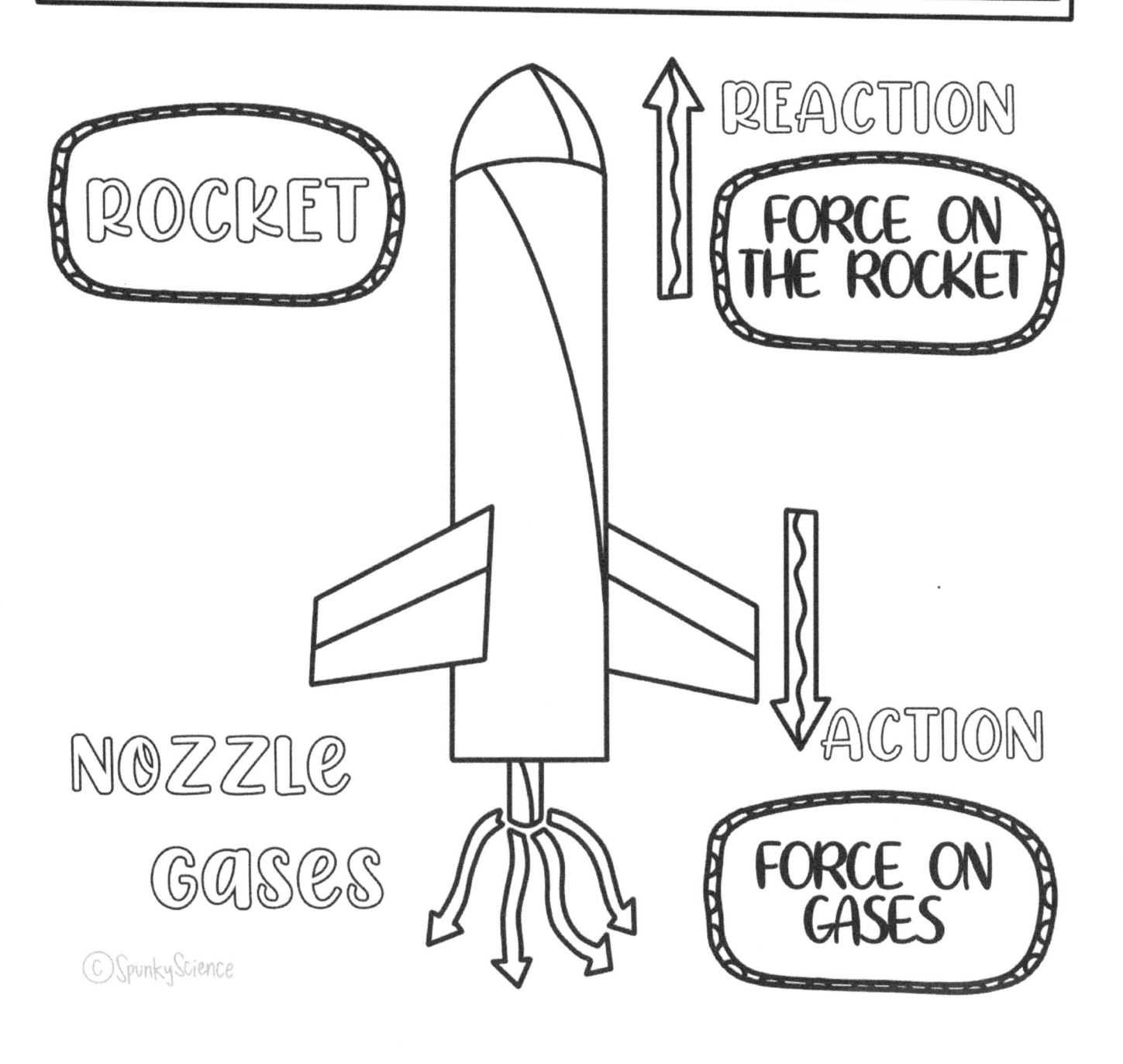

ROCKET

REACTION

FORCE ON THE ROCKET

NOZZLE

GASES

ACTION

FORCE ON GASES

# PHYSICAL CHANGES

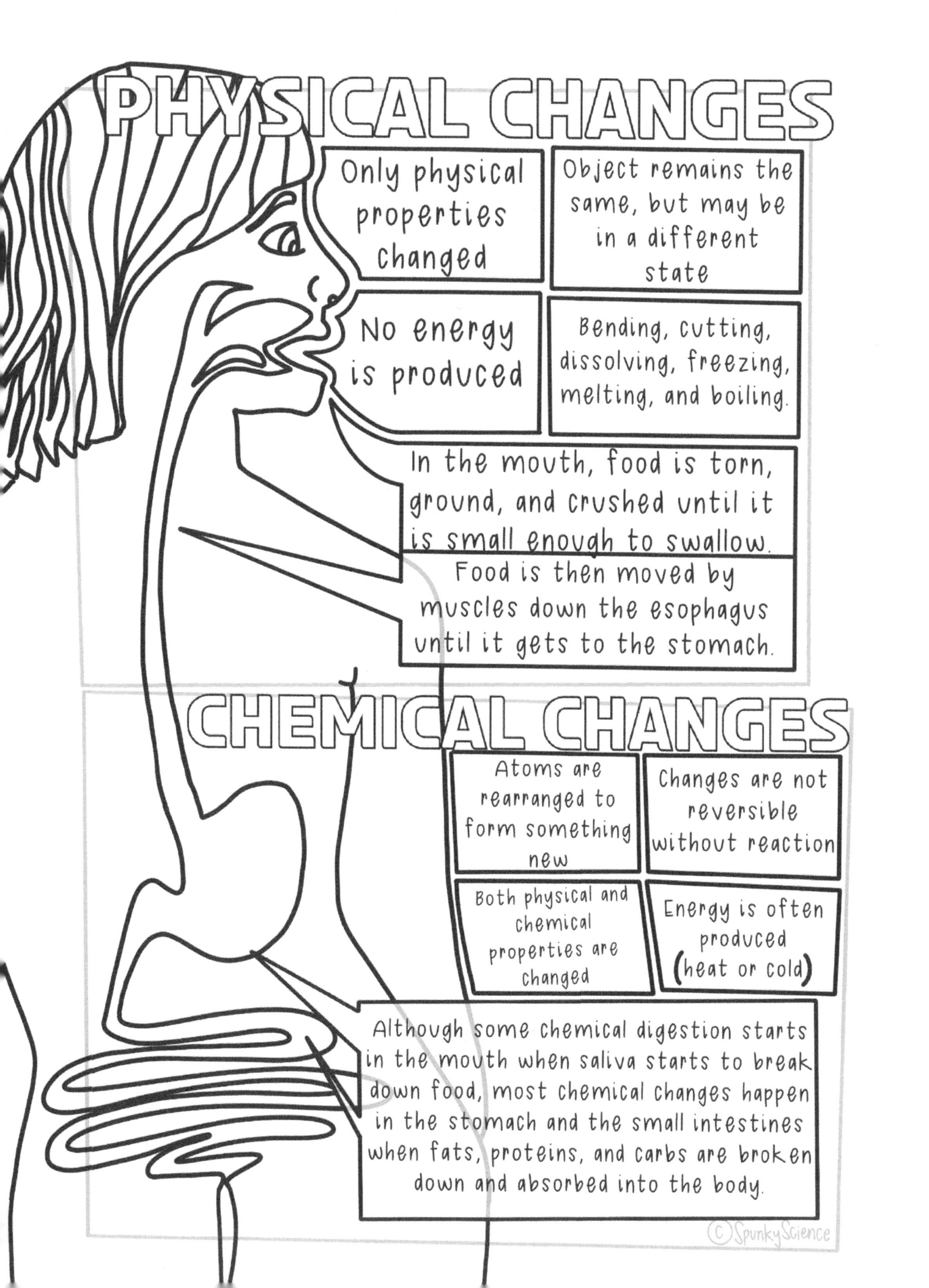

| Only physical properties changed | Object remains the same, but may be in a different state |
| --- | --- |
| No energy is produced | Bending, cutting, dissolving, freezing, melting, and boiling. |

In the mouth, food is torn, ground, and crushed until it is small enough to swallow. Food is then moved by muscles down the esophagus until it gets to the stomach.

# CHEMICAL CHANGES

| Atoms are rearranged to form something new | Changes are not reversible without reaction |
| --- | --- |
| Both physical and chemical properties are changed | Energy is often produced (heat or cold) |

Although some chemical digestion starts in the mouth when saliva starts to break down food, most chemical changes happen in the stomach and the small intestines when fats, proteins, and carbs are broken down and absorbed into the body.

© SpunkyScience

# STATES of MATTER
## EXPLAINED

**BEST ICED COFFEE CO.**

**SOLID:**
- Ice is a solid
- Tightly packed particles
- Keeps its shape even when removed from its container

**GAS:**
- Water vapor is a gas
- Atoms are spread out
- Can be compressed

**LIQUID**
- Coffee is a liquid
- It takes the shape of the container

**PLASMA:**
Plasma is super heated matter. So hot that the electrons are ripped away from the atoms.

# WHAT IS DENSITY?

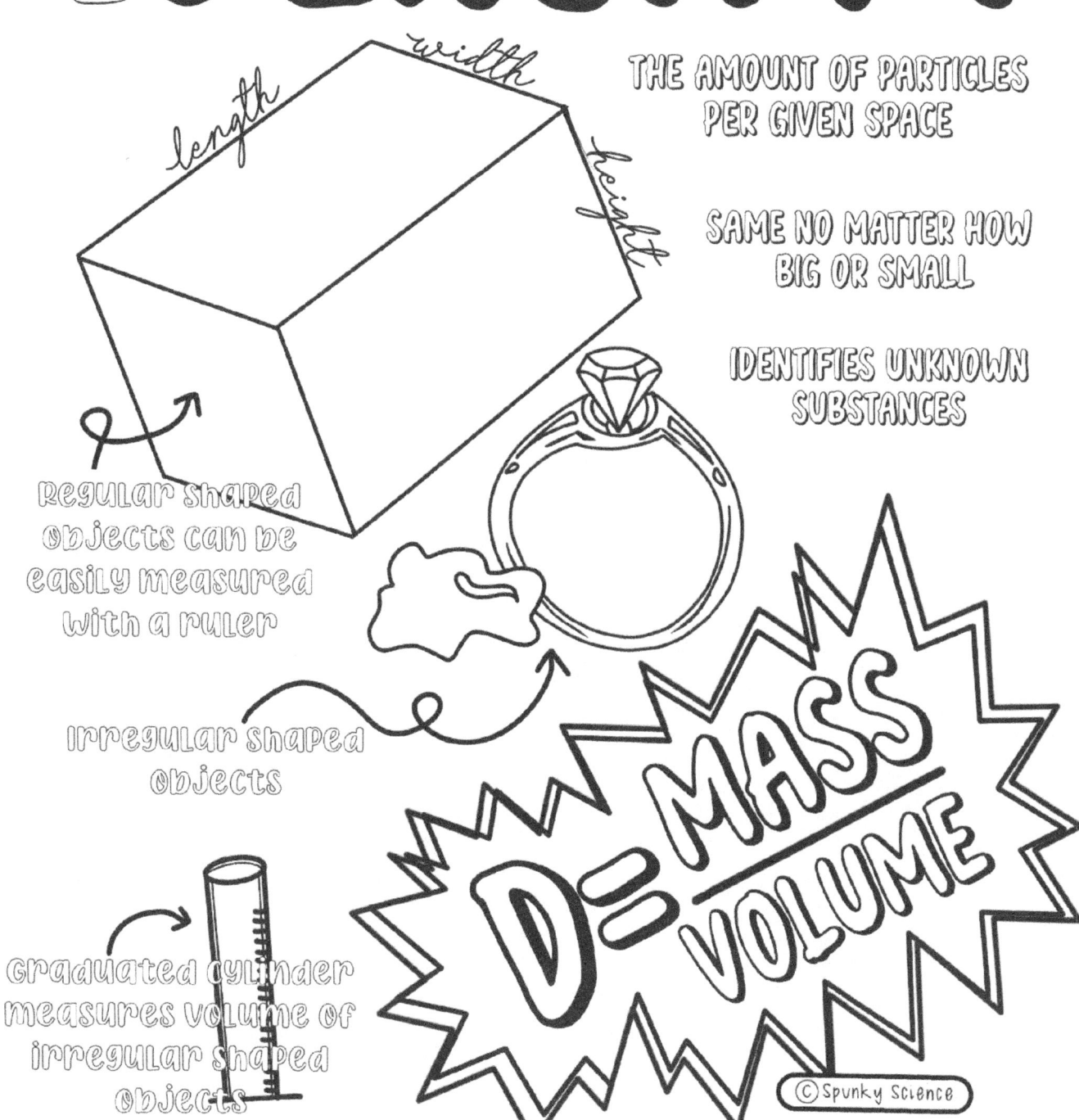

length, width, height

THE AMOUNT OF PARTICLES PER GIVEN SPACE

SAME NO MATTER HOW BIG OR SMALL

IDENTIFIES UNKNOWN SUBSTANCES

Regular shaped objects can be easily measured with a ruler

Irregular shaped objects

Graduated cylinder measures volume of irregular shaped objects

$$D = \frac{MASS}{VOLUME}$$

© Spunky Science

# Signs of a Chemical Change

**TEMPERATURE CHANGE** — HOT OR COLD — fireworks

**FORMS A PRECIPITATE** — Solid

**UNEXPECTED COLOR CHANGE** — ex: clock reactions and pH tests

**GAS CREATED** — Fizz! Pop!

**ODOR OR SMELL PRODUCED**

© Spunky Science

# KINETIC ENERGY

**Ex:** An airplane moving at its fastest speed

Energy due to motion

- Can't sit still! **0% KE**
- Fastest speed **100 KE**
- Slowed down and increased PE. **50% KE**

© Spunky Science

# POTENTIAL ENERGY

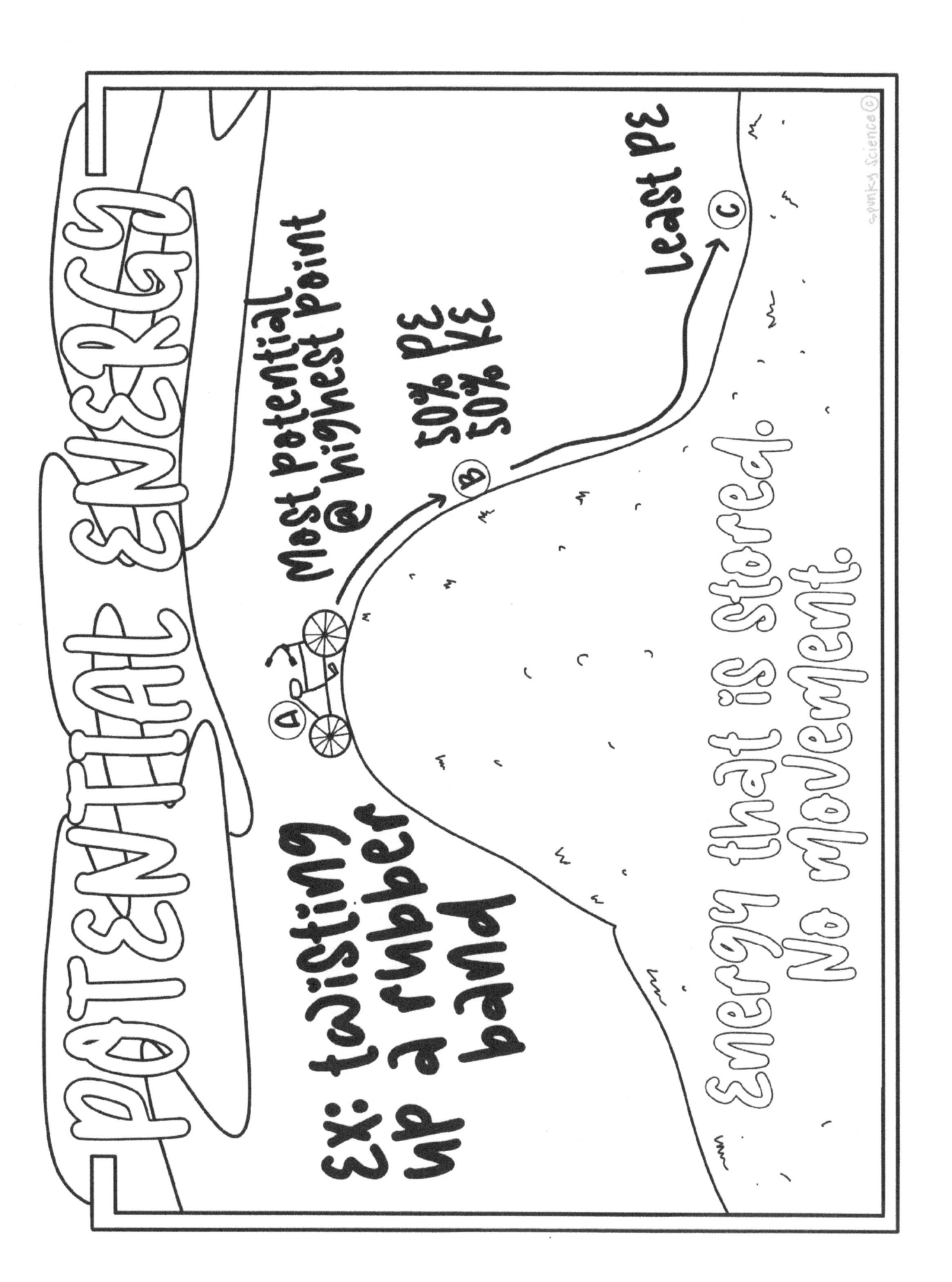

Ex: twisting up a rubber band

Most potential @ highest point

- A
- B — 50% PE / 50% KE
- C — Least PE

Energy that is stored. No movement.

# PRISM

WHITE LIGHT

R
O
Y
G
B
I
V

SNELL'S LAW

INDEX OF REFRACTION

DISPERSING LIGHT

ISSAC NEWTON

BROKEN INTO WAVELENGTHS

RAINBOW

TRIANGULAR PRISM

spunky Science ©

# PERIODIC TABLE OF ELEMENTS

## Fast Facts
→ Each column is called a group
→ The elements in each group have the same number of electrons in their outer orbital
→ Each row is called a period.
→ 90 elements on the PT are natural while the rest are man made.

## COLOR KEY:
- ALKALI METAL
- ALKALINE EARTH METAL
- METALLOID
- TRANSITION METAL
- METAL
- NONMETAL
- NOBLE GAS
- LANTHANIDES
- ACTINIDES
- HALOGEN

# REACTIVITY

The reactivity of an element is how easily it reacts with another substance. The level of reactivity depends on the number of valence electrons.

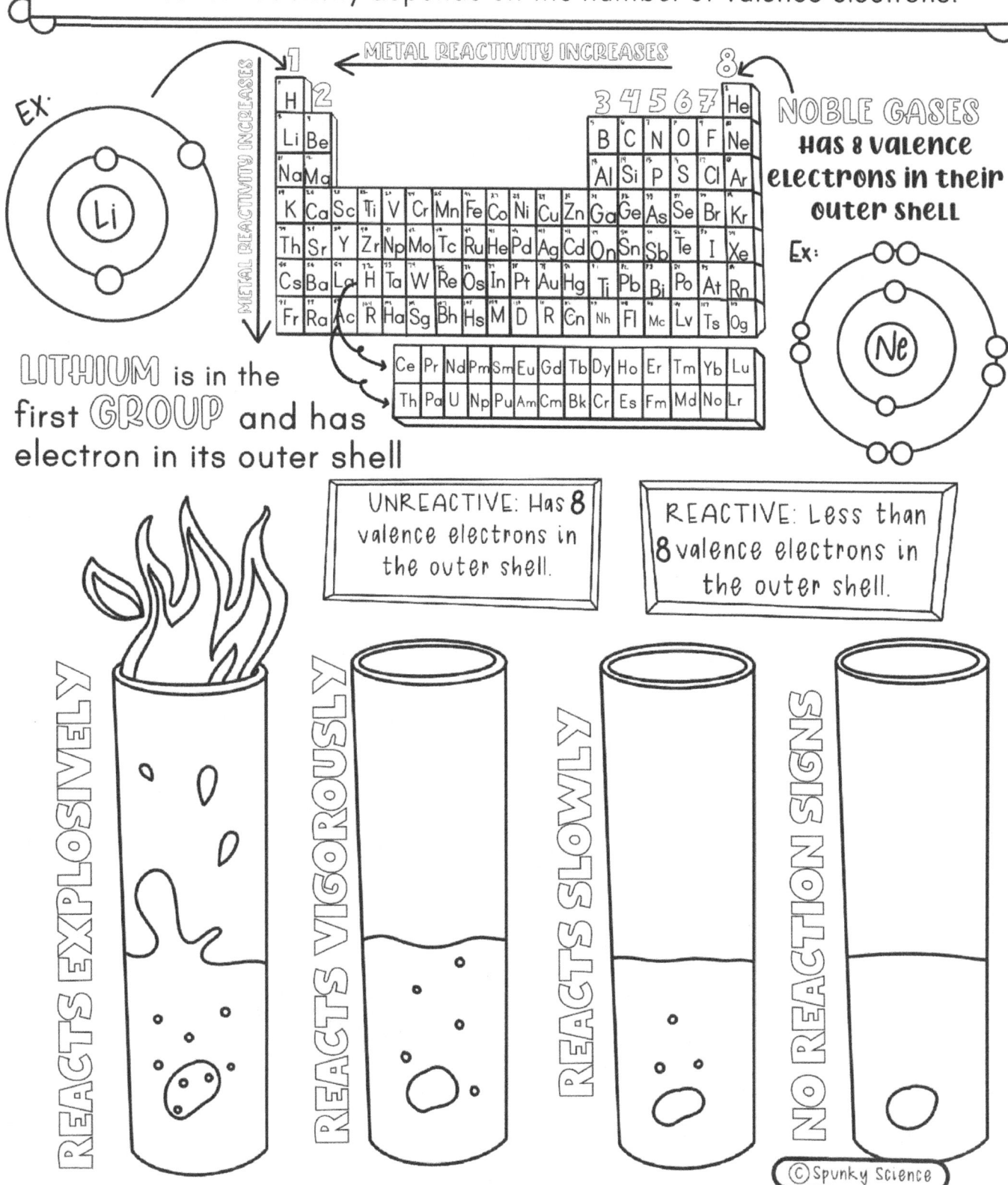

EX:

← METAL REACTIVITY INCREASES

METAL REACTIVITY INCREASES ↓

NOBLE GASES
Has 8 valence electrons in their outer shell

Ex:

LITHIUM is in the first GROUP and has ___ electron in its outer shell

UNREACTIVE: Has 8 valence electrons in the outer shell.

REACTIVE: Less than 8 valence electrons in the outer shell.

REACTS EXPLOSIVELY

REACTS VIGOROUSLY

REACTS SLOWLY

NO REACTION SIGNS

© Spunky Science

Made in the USA
Las Vegas, NV
03 August 2025